T0327489

Distributed Generation

Distributed Generation
Induction and Permanent Magnet Generators

LOI LEI LAI

City University, London, UK

TZE FUN CHAN

Hong Kong Polytechnic University, China

IEEE PRESS

John Wiley & Sons, Ltd

Other Wiley Editorial Offices

John Wiley & Sons Inc., 111 River Street, Hoboken, NJ 07030, USA

Jossey-Bass, 989 Market Street, San Francisco, CA 94103-1741, USA

Wiley-VCH Verlag GmbH, Boschstr. 12, D-69469 Weinheim, Germany

John Wiley & Sons Australia Ltd, 42 McDougall Street, Milton, Queensland 4064, Australia

John Wiley & Sons (Asia) Pte Ltd, 2 Clementi Loop #02-01, Jin Xing Distripark, Singapore 129809

John Wiley & Sons Canada Ltd, 6045 Freement Blud, Mississauga, ONT L5R 4J3

Wiley also publishes its books in a variety of electronic formats. Some content that appears in print may not be available in electronic books.

Anniversary Logo Design: Richard J. Pacifico

British Library Cataloguing in Publication Data

A catalogue record for this book is available from the British Library

ISBN- 978-0470-06208-1

Typeset in 11/13pt Times by Aptara Inc., India.

This book is printed on acid-free paper responsibly manufactured from sustainable forestry in which at least two trees are planted for each one used for paper production.

Contents

Distributed Generation: Induction and Permanent Magnet Generators   L. L. Lai and T. F. Chan
© 2007 John Wiley & Sons, Ltd

Foreword

The concerns over the effects of global warming have seen considerable emphasis being placed on the need to reduce emissions from large, central, fossil-fired power stations. As a result there has been a renaissance in distributed power generating devices and their effects on associated transmission systems. The authors are to be congratulated on producing a comprehensive review of all aspects of induction and permanent magnet generators used in distributed power generation systems. They address the detailed aspects of analysis of these systems, from theory to practical implementation for the latest technologies. This book provides the background and tools for the readers to develop distributed power plants and understand the applications and deployment of induction generator and permanent magnet systems under deregulated power regimes.

This book will be an invaluable aid for engineers, consultants, regulators and environmentalists involved in energy production and its delivery, helping them evaluate renewable energy sources and integrate them into an efficient energy delivery system. Designers, operators and planners will likewise appreciate its unique contribution to the professional literature. Researchers in energy policy as well as electrical engineering students will find that the issues raised in distributed generation, based on induction or permanent magnet generators, show that this is an exciting field in the industry and a necessary area for further research. The book will make a strategic impact on sustainable development in energy, promote environmental impact and improve competitiveness of green energy.

Distributed Generation: Induction and Permanent Magnet Generators L. L. Lai and T. F. Chan
© 2007 John Wiley & Sons, Ltd

Robert Hawley CBE, DSc, FREng, FRSE, HonFIET
IEE President 1996–1997
Past Chairman of the Engineering and Technology Board
Past Chairman of the Particle Physics and Astronomy Research Council
Vice-Chancellor of The World Nuclear University

Preface

Conservation of energy resources, environmental protection and sustainable development are the three major challenges that the world faces at present. One important issue is to satisfy the energy needs of people without causing rapid depletion of the natural energy resources and degradation of the environment. A general consensus among countries of the world is that greater emphasis should be placed on the use of renewable energy resources for electric power generation. Many developing countries have abundant renewable energy resources, but these resources are invariably located in remote regions, thereby creating a number of obstacles for their deployment. The problem can readily be solved if the region is already served by a three-phase grid.

This book presents a comprehensive exposition of induction and permanent magnet generators for distributed power generation, a technology that could help to enable efficient, renewable energy production in both the developed and developing world. The authors examine in detail the theory behind these generators, as well as practical design aspects and implementation issues for the latest technologies. The book starts with a full introduction to distributed energy generation, outlining the reasons for developing this technology and its potential economic impact. There is also a useful overview of the different types of generators that can be used. The following chapters discuss various types of induction generators, including the modelling and testing of these in conjunction with distributed power systems, and the issues of voltage and frequency control. The penultimate chapter analyses permanent magnet generators, looking at the construction of permanent magnet synchronous generators with an inset rotor.

Distributed Generation: Induction and Permanent Magnet Generators   L. L. Lai and T. F. Chan
© 2007 John Wiley & Sons, Ltd

Chapter 1 discusses distributed generation (DG). Depending on the operating scheme and relative performance of the DG system and the power plants supplying the grid, fuel consumption, carbon and other pollutant emissions, and noise pollution can all increase or decrease with DG adoption. For these reasons, DG policy needs to encourage applications that benefit the public, while discouraging those from which the public incurs a net cost. Inherent in this, there is a need to analyse DG costs and benefits and the influence of public policy on DG adoption and operation. While DG may itself become a dominant force in the provision of energy, it is its capability to be used in numerous locations and become integrated into the grid that gives it its greatest value.

Chapter 2 contains a detailed discussion on generator applications. Permanent magnet (PM) synchronous machines are suitable for distributed power generation applications. Advances in PM material technology have stimulated the development of more efficient and more compact generator units. Asynchronous (or induction) generators can also be used to generate alternating current. The induction generator (IG) is very reliable, and tends to be comparatively inexpensive. This type of generator is widely used in wind energy conversion systems, as well as in some small hydro power plants. The use of IGs in grid or isolated power systems is discussed. Various aspects regarding doubly fed IGs, such as their operation, control, performance, impact on power quality, and wind energy production, are briefly reviewed.

In Chapter 3, the general principle of phase balancing for a three-phase IG operating on a single-phase power system is investigated and several practical phase balancing schemes are proposed, including those that involve dissipative elements and current injection transformers. The feasibility of phase balancing for a three-phase IG operating on a single-phase power system has been investigated. A general analysis for the IG with phase converters is presented, and expressions for determination of the phase converter elements are given. The effects of phase balancing on the output power, system power factor and efficiency are also discussed. It is also demonstrated that satisfactory generator performance is obtained with fixed phase converter elements if the speed variation is limited by turbine control. The theoretical analysis is verified by experiments on a small induction machine. The feasibility of a grid-connected single-phase IG system based on the Smith connection is also demonstrated. A systematic analysis using the method of symmetrical components is presented for evaluation of the generator performance at different rotor speeds. The effect of capacitances on the generator performance is investigated too. From a consideration of the voltage unbalance factor, a simple dual-mode control strategy that gives satisfactory generator performance over a wide range of power output is proposed. The theoretical analysis is validated by experiments performed on a small induction machine. Microcontroller-based multi-mode operation of a grid-connected single-phase IG with the Smith connection is presented. The principle and performance analysis of Smith's Mode C circuit are included. This enables a three-phase IG to be operated satisfactorily on a single-phase grid. The advantageous features of this phase balancing scheme include simplicity of circuit configuration, low cost, high efficiency and excellent system power factor.

Chapter 4 presents a new approach for analysing the performance of a grid-connected single-phase IG with the Steinmetz connection. A coupled circuit and field approach based on the two-dimensional finite element method is adopted in order to account for the asymmetrical stator winding connection as well as the complex magnetic field in the machine and the distribution of current in the rotor winding. Since the proposed method is based on rigorous machine modelling and is very general, it can be applied to other asymmetrical IG configurations.

Chapter 5 discusses the various circuit configurations for self-excited induction generators (SEIGs) used in autonomous (also known as stand-alone or isolated) power systems. The analysis of a three-phase SEIG with the Steinmetz connection is presented. The effects of load impedance, speed and power factor on the capacitance requirement are studied. The theoretical analysis is verified by experiments on a small induction machine. The steady-state performance of a single-phase self-regulated SEIG (SRSEIG) using a three-phase machine is analysed. Other advantages of the single-phase SRSEIG include good winding utilization, large power output, high efficiency and a small voltage regulation. The conditions for achieving perfect phase balance in the three-phase machine, which supplies single-phase loads, are also deduced from symmetrical components analysis and the phasor diagram. Since the circuit configuration of the proposed single-phase SRSEIG is extremely simple and only static capacitors are required, the generator can be conveniently implemented for use in low-cost, single-phase, autonomous power generation schemes. The principle and analysis of a novel excitation scheme for a stand-alone three-phase IG that supplies single-phase loads, namely the SMSEIG, is also presented. By adopting the Smith connection with appropriate values of excitation capacitances, balanced operation of the three-phase machine can be achieved. The experimental investigations described confirm the feasibility of the proposed excitation scheme. The SMSEIG has the advantages of low cost, high efficiency and large power output, and as such is an economical choice when developing autonomous single-phase power systems in remote regions.

Chapter 6 presents the voltage and frequency control of a self-excited slip-ring IG (SESRIG) by varying the external rotor resistance. Steady-state performance and the control characteristics of the SESRIG are obtained from an equivalent circuit analysis. With constant load impedance and excitation capacitance, both the frequency and the output voltage of the SESRIG can be maintained constant by rotor resistance control over a wide range of speed without exceeding the stator current limit. A properly tuned PI controller enables good steady-state accuracy and satisfactory dynamic response to be obtained on the generator system. The proposed scheme could be used in a low-cost, variable speed, wind energy system for providing good-quality electric power to remote regions.

Chapter 7 presents the analysis and performance of a three-phase synchronous generator with inset PM rotor. It is demonstrated that the voltage regulation is significantly improved as a result of the inverse saliency feature of the inset PM rotor construction. Experiments performed on a prototype generator confirm the accuracy of the theoretical analysis, and relevant equations for lagging-power-factor loads are developed based on the two-axis model. The conditions for achieving zero voltage regulation, extremum points on the load characteristic and maximum power output are deduced analytically. A method for computing the d-axis and q-axis synchronous reactances for the PMSG are presented. Magnetic saturation, including the effect of rotor magnets, is covered in the analysis. A coupled circuit and field approach is adopted for the performance analysis of a three-phase PMSG with inset rotor and the two-dimensional finite element method is used for the field solution. The solver developed also accounts for the effect of saturation on the air gap flux density and the load characteristics. The experimental voltage and current waveforms are similar to those computed from the 2-D FEM.

Chapter 8 concludes with a consideration of future trends and appendices containing technical data on experimental machines are included.

This book provides invaluable information for power engineers, educators, system operators, managers, planners and researchers.

Loi Lei Lai and Tze Fun Chan
May, 2007

Acknowledgements

The authors wish to thank Wiley in supporting this project.

The authors are indebted to Mr. Lie-Tong Yan, retired professor of Electrical Engineering Department, Tsinghua University, Beijing China, for his contributions to finite element analysis in Chapters 4 and 7.

The second author wishes to thank the Department of Electrical Engineering, the Hong Kong Polytechnic University, Hong Kong, China for the research facilities and support provided.

The authors thank Dr Harald Braun, Research Fellow in Energy Systems Group of City University to produce some of the drawings. The arrangement of the index by Miss Qi Ling Lai and Chun Sing Lai is much appreciated.

The authors would also like to thank the IEEE for permission to reproduce the following:

Figures 2.1–2.3 from T.F. Chan and L.L. Lai, 'Permanent-magnet machines for distributed power generation: a review', Paper number 07GM0593, IEEE 2007 General Meeting, Tampa, FL, USA, 24–28 June 2007.

Figures 3.12–3.13, 3.15–3.27 and Tables 3.4–3.5 from T.F. Chan and L.L. Lai, 'Single-phase operation of a three-phase induction generator with the Smith connection', *IEEE Transactions on Energy Conversion*, Vol. 17, No. 1, March 2002, 47–54.

Figures 4.1–4.14 and Tables 4.1–4.2 from T.F. Chan, L.L. Lai and L.-T. Yan, 'Finite element analysis of a single-phase grid-connected induction generator with the Steinmetz connection', *IEEE Transactions on Energy Conversion*, Vol. 18, No. 2, June 2003, 321–329.

Figures 5.4–5.11 from T.F. Chan and L.L. Lai, 'Capacitance requirements of a three-phase induction generator self-excited with a single capacitance and supplying a single-phase load', *IEEE Transactions on Energy Conversion*, Vol. 17, No. 1, March 2002, 90–94.

Figures 6.9–6.13 and Table 6.1 from T.F. Chan, K.A. Nigim and L.L. Lai, 'Voltage and frequency control of self-excited slip-ring induction generators', *IEEE Transactions on Energy Conversion*, Vol. 19, No. 1, March 2004, 81–87.

Figures 7.1–7.9 and 7.22–7.23 from T.F. Chan, L.L. Lai and L.-T. Yan, 'Performance of a three-phase A.C. generator with inset NdFeB permanent-magnet rotor', *IEEE Transactions on Energy Conversion*, Vol. 19, No. 1, March 2004, 88–94.

And the IET for permission to reproduce the following:

Figures 7.10–7.17, 7.19–21 and 7.24 from T.F. Chan and L.L. Lai, 'Permanent-magnet synchronous generator with inset rotor for autonomous power system applications', *IEE Proceedings – Generation, Transmission & Distribution*, Vol. 151, No. 5, September 2004, 597–603.

Figures 7.25–7.33 and 7.35–7.36 from T.F. Chan, L.L. Lai and L.-T. Yan, 'Analysis of a stand-alone permanent-magnet synchronous generator using a time-stepping coupled field-circuit method', *IEE Proceedings – Electric Power Applications*, Vol. 152, No. 6, November 2005, 1459–1467.

Last but not least, thanks to John Wiley & Sons, Ltd for supporting the preparation of this book and for the extremely pleasant cooperation.

About the Authors

Loi Lei Lai graduated from Aston University in Birmingham with a BSc and a PhD. He was awarded a DSc by City University London. He is also an honorary graduate of City University. In 1984, he was appointed Senior Lecturer at Staffordshire Polytechnic. From 1986 to 1987, he was a Royal Academy of Engineering Industrial Fellow to both GEC Alsthom Turbine Generators Ltd and the Engineering Research Centre. He is regularly involved in consultancy, e.g. in 1994 he carried out a study on the Channel Tunnel power system and gave advice on its protection. He is currently Head of Energy Systems Group and Chair in Electrical Engineering at City University London. In the last decade, Professor Lai has authored/co-authored 200 technical publications. He has also written a book entitled *Intelligent System Applications in Power Engineering – Evolutionary Programming and Neural Networks* and, in 2001, edited the book *Power System Restructuring and Deregulation – Trading, Performance and Information Technology*, both published by John Wiley & Sons, Ltd. He was awarded the IEEE Third Millennium Medal and won the IEEE Power Engineering Society, United Kingdom and Republic of Ireland (UKRI) Chapter, Outstanding Engineer Award in 2003. In 1995, he received a high-quality paper prize from the International Association of Desalination, USA, and in 2006 he was awarded a Prize Paper by the IEEE Power Engineering Society Energy Development and Power Generation Committee. He is a Fellow of the IEEE and the IET (Institution of Engineering and Technology).

Among his professional activities, he is a Founder and was the Conference Chairman of the International Conference on Power Utility Deregulation, Restructuring and Power Technologies (DRPT) 2000, co-sponsored by the IEE (now IET) and

Distributed Generation: Induction and Permanent Magnet Generators   L. L. Lai and T. F. Chan
© 2007 John Wiley & Sons, Ltd

IEEE. He reviews grant proposals regularly for the EPSRC, Australian Research Council and Hong Kong Research Grant Council. In 2001, he was invited by the Hong Kong Institution of Engineers to be Chairman of an Accreditation Visit Team to accredit the BEng (Hons) degree in Electrical Engineering. Since 2005, Professor Lai has been invited as a judge for the Power/Energy Category in the IET Innovation in Engineering Awards. He was also Student Recruitment Officer of the IEEE UKRI Section Executive Committee. He is a member of the Intelligent Systems Subcommittee in Power System Analysis, Computing and Economics Committee, IEEE Power Engineering Society; a Member of the Executive Team of the Power Trading and Control Technical and Professional Network, IET; an Editor of the *IEE Proceedings – Generation, Distribution and Generation* (now *IET Generation, Distribution and Generation*); an Editorial Board Member of the *International Journal of Electrical Power & Energy Systems* published by Elsevier Science Ltd, UK; International Advisor, *Hong Kong Institution of Engineers (HKIE) Transactions*; and an Editorial Board Member of the *European Transactions on Electrical Power* published by Jhon Wiley & Sons, Ltd. He was a Research Professor at Tokyo Metropolitan University, is also Visiting Professor at Southeast University Nanjing and Guest Professor at Fudan University, Shanghai. He has also been invited to deliver keynote addresses and plenary speeches to several major international conferences sponsored by the IET and IEEE.

Tze Fun Chan received his BSc (Eng) and MPhil degrees in electrical engineering from the University of Hong Kong in 1974 and 1980, respectively. He received his PhD in electrical engineering from City University London in 2005. Currently, Dr Chan is an Associate Professor in the Department of Electrical Engineering, Hong Kong Polytechnic University, where he has been since 1978. His research interests are self-excited AC generators, brushless AC generators and permanent magnet machines. In June 2006, he was awarded a Prize Paper by the IEEE Power Engineering Society Energy Development and Power Generation Committee.

1

Distributed Generation

1.1 Introduction

Distributed generation (DG) is related with the use of small generating units installed at strategic points of the electric power system or locations of load centres [1]. DG can be used in an isolated way, supplying the consumer's local demand, or integrated into the grid supplying energy to the remainder of the electric power system. DG technologies can run on renewable energy resources, fossil fuels or waste heat. Equipment ranges in size from less than a kilowatt (kW) to tens of megawatts (MW). DG can meet all or part of a customer's power needs. If connected to a distribution or transmission system, power can be sold to the utility or a third party.

   DG and renewable energy sources (RES) have attracted a lot of attention worldwide [2–4]. Both are considered to be important in improving the security of energy supplies by decreasing the dependency on imported fossil fuels and in reducing the emissions of greenhouse gases (GHGs). The viability of DG and RES depends largely on regulations and stimulation measures which are a matter of political decisions.

1.2 Reasons for DG

DG can be applied in many ways and some examples are listed below:

- It may be more economic than running a power line to remote locations.
- It provides primary power, with the utility providing backup and supplemental power.
- It can provide backup power during utility system outages, for facilities requiring uninterrupted service.

Distributed Generation: Induction and Permanent Magnet Generators   L. L. Lai and T. F. Chan
© 2007 John Wiley & Sons, Ltd

- For cogeneration, where waste heat can be used for heating, cooling or steam. Traditional uses include large industrial facilities with high steam and power demands, such as universities and hospitals.
- It can provide higher power quality for electronic equipment.
- For reactive supply and voltage control of generation by injecting and absorbing reactive power to control grid voltage.
- For network stability in using fast-response equipment to maintain a secure transmission system.
- For system black-start to start generation and restore a portion of the utility system without outside support after a system collapse.

DG can provide benefits for consumers as well as for utilities. Some examples are listed below:

- Transmission costs are reduced because the generators are closer to the load and smaller plants reduce construction time and investment cost.
- Technologies such as micro turbines, fuel cells and photovoltaics can serve in several capacities including backup or emergency power, peak shaving or base load power.
- Given the uncertainties of power utility restructuring and volatility of natural gas prices, power from a DG unit may be less expensive than conventional electric plant. The enhanced efficiency of combined heat and power (CHP) also contributes to cost savings [5].
- DG is less capital intensive and can be up and running in a fraction of the time necessary for the construction of large central generating stations.
- Certain types of DG, such as those run on renewable resources or clearer energy systems, can dramatically reduce emissions as compared with conventional centralized large power plants.
- DG reduces the exposure of critical energy infrastructure to the threat of terrorism.
- DG is well suited to providing the ancillary services necessary for the stability of the electrical system.
- DG is most economical in applications where it covers the base load electricity and uses utility electricity to cover peak consumption and the load during DG equipment outages, i.e. as a standby service.
- DG can offset or delay the need for building more central power plants or increasing transmission and distribution infrastructure, and can also reduce grid congestion, translating into lower electricity rates for all utility customers.
- Smaller, more modular units require less project capital and less lead-time than large power plants. This reduces a variety of risks to utilities, including forecasting of load/resource balance and fuel prices, technological obsolescence and regulatory risk.
- DG can provide the very high reliability and power quality that some businesses need, particularly when combined with energy storage and power quality technologies.

- Small generating equipment can more readily be resold or moved to a better location.
- DG maximizes energy efficiency by enabling tailored solutions for specific customer needs such as combined heat and power systems.
- By generating power at or very near the point of consumption where there is congestion, DG can increase the effective transmission and distribution network capacity for other customers.
- DG can reduce customer demands from the grid during high demand periods.
- DG can provide very high-quality power that reduces or eliminates grid voltage variation and harmonics that negatively affect a customer's sensitive loads.
- DG may allow customers to sell excess power or ancillary services to power markets, thus increasing the number of suppliers selling energy and increasing competition and reducing market power.
- DG can reduce reactive power consumption and improve voltage stability of the distribution system at lower cost than voltage-regulating equipment.
- DG eliminates the need for costly installation of new transmission lines, which frequently have an environmental issue.
- DG reduces energy delivery losses resulting in the conservation of vital energy resources.
- DG expands the use of renewable resources, such as biomass cogeneration in the paper industry, rooftop solar photovoltaic systems on homes, and windmills further to improve energy resource conservation.
- DG offers grid benefits like reduced line loss and increased reliability [6]. From a grid security standpoint, many small generators are collectively more reliable than a few big ones. They can be repaired more quickly and the consequences of a small unit's failure are less catastrophic. DG eliminates potential blackouts caused by utilities' reduced margin of generation reserve capacity. Figure 1.1 shows the number of customers affected by major blackouts during the last four decades worldwide.

1.3 Technical Impacts of DG

DG technologies include engines, small wind turbines, fuel cells and photovoltaic systems. Despite their small size, DG technologies are having a stronger impact in electricity markets. In some markets, DG is actually replacing the more costly grid electricity. However, there are technical issues that deserve attention.

1.3.1 DG Technologies

No single DG technology can accurately represent the full range of capabilities and applications or the scope of benefits and costs associated with DG. Some of these technologies have been used for many years, especially reciprocating engines and gas turbines. Others, such as fuel cells and micro turbines, are relative

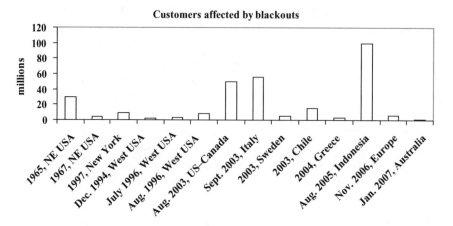

Figure 1.1   *Number of customers affected by major blackouts. Reproduced by permission of T.-F. Chan and L. L. Lai, 'Permanent-Magnet Machines for Distributed Power Generation: A Review' paper No. 07GM0593, IEEE 2007 General Meeting, Tampa, USA. Copyright © (2007) IEEE + reference to publication, author*

new developments. Several DG technologies are now commercially available, and some are expected to be introduced or substantially improved within the next few years [7].

Reciprocating engines. Diesel and gas reciprocating engines are well-established commercial DG technologies. Industrial-sized diesel engines can achieve fuel efficiencies in excess of 40 % and are relatively low cost per kilowatt. While nearly half of the capacity was ordered for standby use, the demand for units for continuous or peak use has also been increasing.

Gas turbines. Originally developed for jet engines, gas turbines are now widely used in the power industry. Small industrial gas turbines of 1–20 MW are commonly used in combined heat and power applications. They are particularly useful when higher temperature steam is required than can be produced by a reciprocating engine. The maintenance cost is slightly lower than for reciprocating engines, but so is the electrical conversion efficiency. Gas turbines can be noisy. Emissions are somewhat lower than for engines, and cost-effective NO_x emission control technology is commercially available.

Micro turbines. Micro turbines extend gas turbine technology to units of small size. The technology was originally developed for transportation applications, but is now finding a place in power generation. One of the most striking technical characteristics of micro turbines is their extremely high rotational speed. The turbine rotates up to 120 000 r/min and the generator up to 40 000 r/min. Individual units range from 30 to 200 kW but can be combined into systems of multiple units. Low combustion temperatures can assure very low NO_x emission levels. These turbines make much less noise than an engine of comparable size. Natural gas is expected to be the most common fuel but flare gas, landfill gas or biogas can also be used. The main disadvantages of micro turbines are their short track record and high costs compared with gas engines.

Fuel cells. Fuel cells are compact, quiet power generators that use hydrogen and oxygen to make electricity. The transportation sector is the major potential market for fuel cells, and car manufacturers are making substantial investments in research and development. Power generation, however, is seen as a market in which fuel cells could be commercialized much more quickly. Fuel cells can convert fuels to electricity at very high efficiencies (35–60 %), compared with conventional technologies [8]. As there is no combustion, other noxious emissions are low. Fuel cells can operate with very high reliability and so could supplement or replace grid-based electricity. Only one fuel cell technology for power plants, a phosphoric acid fuel cell plant (PAFC), is currently commercially available. Three other types of fuel cells, namely molten carbonate (MCFC), proton exchange membrane (PEMFC) and solid oxide (SOFC), are the focus of intensive research and development.

Photovoltaic systems. Photovoltaic systems are a capital-intensive, renewable technology with very low operating costs. They generate no heat and are inherently small scale. These characteristics suggest that photovoltaic systems are best suited to household or small commercial applications, where power prices on the grid are highest. Operating costs are very low, as there are no fuelling costs.

Wind. Wind generation is rapidly gaining a share in electricity supply worldwide. Wind power is sometimes considered to be DG, because the size and location of some wind farms make it suitable for connection at distribution voltages.

1.3.2 Thermal Issues

When DG is connected to the distribution network, it alters the load pattern. The amount of feeder load demand will eventually result in the feeder becoming fully loaded. It is most likely that increased levels of DG will cause an increase in the overall current flowing in the network, bringing the components in the network closer to their thermal limits. If the thermal limits of the circuit components are likely to be exceeded by the connection of DG, then the potentially affected circuits would need to be replaced with circuits of a higher thermal rating. This would usually take the form of replacement with conductors of a larger cross-sectional area.

1.3.3 Voltage Profile Issues

Voltage profiles along a loaded distribution network feeder are typically such that the voltage level is at maximum close to the distribution network transformer busbar, and the voltage drops along the length of the feeder as a result of the load connected to the feeder. Voltage drop is generally larger on rural networks, which are commonly radial networks with feeders covering long distances with relatively low-current-capacity conductors, especially at the remote ends of the feeders. The distribution transformer, feeding the distribution network, is fitted with a tap-changer, which controls the setting of the busbar voltage. The tap-changer will be set to ensure that, under maximum feeder loads, the voltage drop along a feeder does not result in voltage levels falling below the lower of the statutory voltage limits.

DG along a distribution feeder will usually have the effect of reducing the voltage drop along the feeder, and may lead to a voltage rise at some points which could push the feeder voltage above the statutory voltage limit. Voltage rise is generally more of a problem on rural radial networks than on interconnected or ring networks, as excessive voltage rise can be initiated by relatively small amounts of DG due to the high impedance of the conductors and because these feeders are often operated close to the statutory upper voltage limit to counter the relatively large voltage drop over the length of such feeders. Voltage rise may be reduced by:

1. Constraining the size of DG plant: the level of voltage rise will depend upon the generation level compared with the minimum load demand.
2. Reinforcing the network (initially using larger conductors with a lower impedance).
3. Operating the generator at a leading power factor (i.e. importing VArs from the network), which will reduce overall power flow and hence reduce voltage drop. However, distribution network operators (DNOs) generally require DG plant to operate as close to unity power factor as possible (i.e. negligible import or export of reactive power).
4. Installing shunt reactor banks to draw additional reactive power from the network. DG could also contribute to voltage flicker through sudden variations in the DG output (e.g. variable wind speeds on turbines), start-up of large DG units or interactions between DG and voltage control equipment on the network. Wind turbines with induction generators will cause voltage disturbances when starting, due to the inrush of reactive current required to energize the rotor. The voltage step that will occur when a wind turbine shuts down from full output, perhaps due to high wind speeds, must also be considered. A short-term reduction in the network voltage means that there is not enough energy to supply the connected load. There are two major causes of these voltage dips: namely, sudden connections of large loads or faults on adjacent branches of the network. When DG is connected to a network and is energized, a voltage step may result from the inrush current flowing into the generator or transformer. Step voltages also occur when a generator (or group of generators) is suddenly disconnected from the network, most likely due to a fault.

When large motor loads are suddenly connected to the network, they draw a current, which can be many times larger than the nominal operating current. The supply conductors for the load are designed for nominal operation; therefore this high current can cause an excessive voltage drop in the supply network. Voltage dips caused by large motor loads can be overcome by installing a starter, which limits the starting current but increases the starting time. Another option is to negotiate with the DNO for a low-impedance connection, though this could be an expensive option depending on the local network configuration. Depending on the reaction time of control systems, there are several options to reduce the severity of voltage dips: that is, to increase DG output, to reduce network loads, to utilize energy from storage devices or energize capacitor banks.

1.3.4   Fault-Level Contributions

A fault can occur in many ways on a network due to a downed overhead line or a damaged underground cable. The current that flows into a fault can come from three sources on a distribution network: namely, infeeds from the transmission system, infeeds from distributed generators or infeeds from loads (with induction motors).

The connection of DG causes fault levels close to the point of connection to increase. This increase is caused by an additional fault level from the generator, and can cause the overall fault level to exceed the designed fault level of the distribution equipment. Increased fault levels can be accommodated, or reduced, by either upgrading equipment or reconfiguring distribution networks.

Induction generators contribute very little to root mean square (RMS) break fault levels, as the fault current from the induction generator quickly collapses as the generator loses magnetic excitation due to the loss of grid supply. However, they contribute more to peak fault levels. Synchronous generators contribute less to the initial peak current compared with induction generators but do have a larger steady-state RMS fault contribution. For generators which are connected to the distribution network via power electronics interfaces, it will be quickly disconnected under network fault conditions when a current is 20 % higher than the rated current. As a doubly fed induction generator (DFIG) is only partially connected via power electronics, the RMS break fault current contribution is low. However, the peak current contribution can be up to six times the rated current.

1.3.5   Harmonics and Interactions with Loads

In ideal electricity network the voltage would have a perfectly sinusoidal waveform oscillating, for example, at 50 cycles per second. However, any capacitive or inductive effects, due to switching of devices such as large cables, network reactors, rectified DC power supplies, variable speed motor drives and inverter-coupled generators, will introduce or amplify 'harmonic' components into the voltage sine wave, thereby distorting the voltage waveform. It is expected that small-scale micro wind and solar generation will be inverter connected. Inverter connections incorporate the use of a high proportion of switching components that have the potential to increase harmonic contributions.

1.3.6   Interactions Between Generating Units

Increasing levels of intermittent renewable generation and fluctuating inputs from CHP units will ultimately make it more difficult to manage the balance between supply and demand of the power system. Unless the DG can offer the same control functions as the large generators on the system, the amount of generation reserve required when there is a significant contribution to the system from DG will need to be increased.

1.3.7   Protection Issues

Distribution networks were designed to conduct current from high to low voltages and protection devices are designed to reflect this concept. Under conditions of current flow in the opposite direction, protection mal-operation or failure may occur with consequent increased risk of widespread failure of supply.

Due to opposite current flow, the reach of a relay is shortened, leaving high-impedance faults undetected. When a utility breaker is opened, a portion of the utility system remains energized while isolated from the remainder of the utility system, resulting in injuries to the public and utility personnel. Figure 1.2 shows an islanding situation where IM, SM and CB stand for induction machine, synchronous machine and circuit breaker, respectively [9].

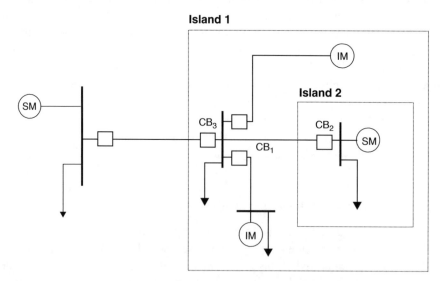

Figure 1.2   *Islanding operation. Reproduced by permission of T.-F. Chan and L. L. Lai, 'Permanent-Magnet Machines for Distributed Power Generation: A Review' paper No. 07GM0593, IEEE 2007 General Meeting, Tampa, USA. Copyright ©(2007) IEEE + reference to publication, author*

1.4 Economic Impact of DG

DG has some economic advantages compared with power from the grid, particularly for on-site power production [10–12]. The possibility of generating and using both heat and power generated in a CHP plant can create additional economic opportunities. DG may also be better positioned to use low-cost fuels such as landfill gas.

The relative prices of retail electricity and fuel costs are critical to the competitiveness of any DG option. This ratio varies greatly from country to country. In Japan, for example, where electricity and natural gas prices are high, DG is attractive only for oil-fired generation. In other countries, where gas is inexpensive compared with electricity, DG can become economically attractive. Many DG technologies can be very flexible in their operation. A DG plant can operate during periods of high electricity prices (peak periods) and then be switched off during low-price periods.

The ease of installation of DG also allows the system capacity to be expanded readily to take advantage of anticipated high prices. Some DG assets are portable. In addition to this technological flexibility, DG may add value to some power systems by delaying the need to upgrade a congested transmission or distribution network, by reducing distribution losses and by providing support or ancillary services to the local distribution network.

CHP is economically attractive for DG because of its higher fuel efficiency and low incremental capital costs for heat-recovery equipment. Domestic-level CHP, so-called 'micro-CHP', is attracting much interest, particularly where it uses external combustion engines and in some cases fuel cells. However, despite the potential for short payback periods, high capital costs for the domestic consumer are a significant barrier to the penetration of these technologies.

The provision of reliable power represents the most important market for DG. Emergency diesel generating capacity in buildings, generally not built to export power to the grid, represents several per cent of total peak demand for electricity. Growing consumer demand for higher quality electricity (e.g. 'five nines' or 99.999 % reliability) requires on-site power production.

Many of these technologies can be more energy efficient and cleaner than central station power plants. Modularity is beneficial when load growth is slow or uncertain.

The smaller size of these technologies can better match gradual increases in utility loads. DG also can reduce demand during peak hours, when power costs are highest and the grid is most congested. If located in constrained areas, DG can reduce the need for distribution and transmission system upgrades. Customers can install DG to cap their electricity costs, sell power, participate in demand response programmes, provide backup power for critical loads and supply premium power to sensitive loads.

The biggest potential market for DG is to supplement power supplied through the transmission and distribution grid. On-site power production reduces transmission and distribution costs for the delivery of electricity. These costs average about 30 % of the total cost of electricity. This share, however, varies according to customer size. For very large customers taking power directly at transmission voltage, the total cost and percentage are much smaller; for a small household consumer, network charges may constitute over 40 % of the price.

Small-scale generation has a few direct cost disadvantages over central generation. First, there is a more limited selection of fuels and technologies to generate electricity – oil, natural gas, wind or photovoltaic systems, and, in certain cases, biomass or waste fuels. Second, the smaller generators used in DG cost more per kilowatt to build than larger plants used in central generation. Third, the costs of fuel delivery are normally higher. Finally, unless run in CHP mode, the smaller plants used in DG operate usually at lower fuel conversion efficiencies than those of larger plants of the same type used in central generation. DG uses a more limited selection of fuels. For photovoltaic systems, operating costs are very low, but high capital costs prevent them from competitive with grid electricity.

1.5   Barriers to DG Development

Cooperation, property ownership, personal consumption and security will change attitudes towards DG technologies and make people welcome them to their homes. There is now evidence of strong interest from a small community willing to pay the premium to enjoy green energy [13].

There is significant regional variation in the use of DG systems. This is largely due to the fact that the potential benefits of DG are greater in some areas than others. In some areas, for example, relatively high electricity rates, reliability concerns and DG-friendly regulatory programmes have encouraged comparatively fast DG development. But in many areas, even where DG could offer benefits, projects are often blocked by market and other barriers. The most commonly cited barrier to DG development is the process of interconnecting to distribution and transmission systems. Other barriers include high capital costs, non-uniform regulatory requirements, lack of experience with DG, and tariff structures [14–16].

The lack of experience with competitive markets often increases risk about the use of unconventional power sources. Customers cannot easily sell power from on-site generation to the utility through a competitive bidding process, to a marketer or to other customers directly. For customers, there is a risk of DG being uneconomical; capital investments under market uncertainty; price volatility for the DG system fuel. There is a concern about the reliability and risks that arise from using unconventional technologies/applications with DG.

Utilities have a considerable economic disincentive to embrace distributed resources. Distribution company profits are directly linked to sales. Utilities' revenues are based on how much power they sell and move over their wires, and they lose sales when customers develop generation on site. Interconnecting with customer-owned DG is not in line with a utility's profit motive. Other barriers to the deployment of DG exist on the customer side. A utility has no obligation to connect DG to its system unless the unit is a qualifying facility. If a utility does choose to interconnect, lengthy case-by-case impact studies and redundant safety equipment can easily spoil the economics of DG. If a customer wants the utility to supply only a portion of the customer's load or provide backup power in case of unit failure, the cost of 'standby' and 'backup' rates can be prohibitive. Exit fees and competitive transition charges associated with switching providers or leaving the grid entirely can be burdensome. And obtaining all the necessary permits can be quite difficult.

1.6   Renewable Sources of Energy

These are the natural energy resources that are inexhaustible: for example, wind, solar, geothermal, biomass and small-hydro generation.

Small-hydro energy. Although the potential for small hydroelectric systems depends on the availability of suitable water flow, where the resource exists it can provide cheap, clean, reliable electricity. Hydroelectric plants convert the kinetic energy of a waterfall into electric energy. The power available in a flow of water depends on the vertical distance the water falls and the volume of the flow of water. The water powers a turbine, and its rotation movement is transferred through a shaft to an electric generator. A hydroelectric installation alters its natural surroundings. The effects on the environment must therefore be evaluated during the planning of the project to avoid problems such as noise or damage to ecosystems.

Wind energy. Wind turbines produce electricity for homes, businesses and utilities. Wind power will continue to prosper as new turbine designs currently under development reduce its costs and make wind turbines economically viable in more and more places. Wind speed varies naturally with the time of day, the season and the height of the turbine above the ground. The energy available from wind is proportional to the cube of its speed. A wind generator is used to convert the power of wind into electricity. Wind generators can be divided into two categories, those with a horizontal axis and those with a vertical axis [17]. The Electric Power Research Institute, USA, has stated that wind power offers utilities pollution-free electricity that is nearly cost-competitive with today's conventional sources. However, one environmental concern about wind power is land use. Modern wind turbine technology has made significant advances over the last 10 years. Today, small wind machines of 5 to 40 kW capacity can supply the normal electrical needs of homes and small industries. Medium-size turbines rated from 100 to 500 kW produce most of the commercially generated electricity.

Biomass. The term biomass refers to the Earth's vegetation and many products that come from it. Some of the commonest biomass fuels are wood, agricultural residues and crops grown for energy. Utilities and commercial and industrial facilities use biomass to produce electricity. According to the World Bank, 50 to 60 % of the energy in the developing countries of Asia, and 70 to 90 % of the energy in the developing countries of Africa, come from biomass, and half the world's population cook with wood. In the USA, Japan and Europe, municipal and agricultural waste are being burned to produce electricity.

Solar energy. Solar thermal electric power plants use various concentrating devices to focus sunlight and achieve the high temperatures necessary to produce steam for power. Flat-plate collectors transfer the heat of the Sun to water either directly or through the use of another fluid and a heat exchanger. The market for photovoltaics is rapidly expanding. Homes can use photovoltaic systems to replace or supplement electric power from the utility. A stand-alone residential system consists of solar panels, a battery to store power for use at night, and an inverter to allow conventional appliances to be powered by solar electricity.

Geothermal. Geothermal energy is heat from the Earth that is used directly as hot water or steam, or used to produce electricity. While high-temperature geothermal sites suitable for electricity production are not widespread, low-temperature sites are found almost everywhere in the world and they can provide heating and cooling for buildings. Geothermal systems are located in areas where the Earth's crust is relatively thin. Drilling into the ground and inserting pipes enable hot water or steam to be brought to the surface. In some applications, this is used to provide direct heating to homes. In other areas, the steam is used to drive a turbine to generate electricity. According to the US Energy Information Agency, geothermal energy has the potential to provide the USA with 12 000 megawatts of electricity by the year 2010, and 49 000 megawatts by 2030. It has the potential to provide up to 80 000 megawatts. Geothermal energy resources are found around the world. As a local and renewable energy resource, geothermal energy can help reduce a nation's dependence on oil and other imported fuels. Geothermal heat pumps (GHPs) are an efficient way to heat and cool buildings. GHPs use the normal temperature of the Earth to heat buildings in winter and cool them in summer. GHPs take advantage of the fact that the temperature of the ground does not vary as much from season to season as the temperature of the air.

1.7   Renewable Energy Economics

Generating electricity from the wind makes economic as well as environmental sense: wind is a free, clean and renewable resource which will never run out. The wind energy industry – designing and making turbines, erecting and running them – is growing fast and is set to expand as the world looks for cleaner and more sustainable ways to generate electricity. Wind turbines are becoming cheaper and more powerful, with larger blade lengths which can utilise more wind and therefore produce more electricity, bringing down the cost of renewable generation.

Making and selling electricity from the wind is no different from any other business. To be economically viable the cost of making electricity has to be less than its selling price. In every country the price of electricity depends not only on the cost of generating it, but also on the many different factors that affect the market, such as energy subsidies and taxes. The cost of generating electricity comprises capital costs (the cost of building the power plant and connecting it to the grid), running costs (such as buying fuel and operation and maintenance) and the cost of financing (how the capital cost is repaid).

With wind energy, and many other renewables, the fuel is free. Therefore once the project has been paid for, the only costs are operation and maintenance and fixed costs, such as land rental. The capital cost is high, between 70 and 90 % of the total for onshore projects. The more electricity the turbines produce, the lower the cost of the electricity. This depends on the power available from the wind. Roughly, the power derived is a function of the cube of the wind speed. Therefore if the wind speed is doubled, its energy content will increase eight fold. Turbines in wind farms must be arranged so that they do not shadow each other.

The cost of electricity generated from the wind, and therefore its final price, is influenced by two main factors, namely technical factors and financial perspective. Technical factors are about wind speed and the nature of the turbines, while financial perspective is related to the rate of return on the capital, and the period of time over which the capital is repaid. Naturally, how quickly investors want their loans repaid and what rate of return they require can affect the feasibility of a wind project; a short repayment period and a high rate of return push up the price of electricity generated. Public authorities and energy planners require the capital to be paid off over the technical lifetime of the wind turbine, e.g. 20 years, whereas the private investor would have to recover the cost of the turbines during the length of the bank loan. The interest rates used by public authorities and energy planners would typically be lower than those used by private investors.

Although the cost varies from country to country, the trend is everywhere the same: that is, wind energy is getting cheaper. The cost is coming down for various reasons. The turbines themselves are getting cheaper as technology improves and the components can be made more economically. The productivity of these newer designs is also better, so more electricity is produced from more cost-effective turbines. The cost of financing is also falling as lenders gain confidence in the technology. Wind power should become even more competitive as the cost of using conventional energy technologies rises.

The economics of wind energy are already strong, despite the young age of the industry. The world market in wind turbines continues to boom: a comprehensive policy package combining the best elements of market-based incentives and technology policy approaches will accelerate the implementation of existing clean, energy-efficient technologies; stimulate the development of renewable domestic energy sources; and promote research and development on efficient new technologies. Investment in efficient, clean energy technologies lowers business costs and boosts the productivity and competitiveness of industry. That means faster economic growth, more jobs and higher wages. An emission trading scheme will encourage innovation and stimulate investment in the lowest cost techniques to reduce GHGs. Low-emission plants are used to offset higher emissions made by others, with the aim that the utility in total meets emission requirements. The Clean Development Mechanism (CDM) is an arrangement under the Kyoto Protocol allowing developed countries with a GHG reduction commitment to invest in emission-reducing projects in developing countries as an alternative to what is generally considered more costly emission reductions in their own countries [18]. This mechanism will also help promote renewable generation. Figure 1.3 shows the price trend of EU emissions trading in 2005.

However, renewable energy technologies will introduce new conflicts. For example, a basic parameter controlling renewable energy supplies is the availability of land. At present world food supply mainly comes from land. There is relatively little land available for other uses, such as biomass production and solar technologies. Population growth demands land. Therefore, future land conflicts will be intensified. Although renewable energy technologies often cause fewer

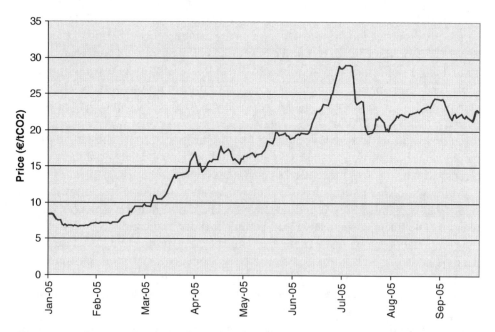

Figure 1.3   *The EU emissions trading scheme. Reproduced by permission of T.-F. Chan and L. L. Lai, 'Permanent-Magnet Machines for Distributed Power Generation: A Review' paper No. 07GM0593, IEEE 2007 General Meeting, Tampa, USA. Copyright © (2007) IEEE + reference to publication, author*

environmental problems than fossil energy systems, they require large amounts of land and therefore compete with agriculture, forestry and other essential land-use systems. Reservoirs constructed for hydroelectric plants have the potential to cause major environmental problems. This water cover represents a major loss of productive agricultural land. Dams may fail, resulting in loss of life and destruction of property. Further, dams alter the existing plant and animal species in an ecosystem, e.g. by blocking fish migration. Generation, transmission and distribution utilities generally plan their systems to meet all of the power needs of all of their customers. They do not encourage their customers to develop on-site generation. In some cases, utilities have actively opposed DG projects.

1.8 Interconnection

1.8.1 Interconnection Standardization

A customer who wants to interconnect DG to the distribution system must undergo a utility's case-by-case interconnection review process [19–24]. Such a process can be time consuming and expensive. Installers thus face higher costs by having to meet interconnection requirements that vary from utility to utility. Additionally, manufacturers are not able to capture the economies of scale in producing package systems with standard safety and power quality protection. The interconnection process would benefit from the pre-certification of specific DG technologies. Recognized, independent or government testing labs (e.g. Underwriters Laboratories) would conduct initial testing and characterization of the safety, power quality and system reliability impacts of DG. They would recommend technical parameters that state legislatures, regulatory agencies or individual utilities could adopt.

1.8.2 Rate Design

The restructuring of electricity markets and an increased reliance on wholesale power purchases have brought distribution into the spotlight. As utilities have divested themselves of generation assets, they have become aware of the importance of distribution services in generating revenue. Usage-based rates help ensure that customers pay the actual costs they impose on the system so that their consumption neither subsidizes nor is subsidized by the consumption of others.

Rates should reflect the grid benefits of DG, like peak shaving, reduced need for system upgrades, capital cost reductions and increased reliability. Standby or backup charges are rates that a customer pays to receive power from the grid at times when its own DG is unavailable. Standby rates are typically based on serving a customer's maximum load at peak demand periods – a worst case scenario which, some argue, should not serve as the basis for rate making. Buyback rates are the prices a utility pays for excess generation from a customer's own DG unit. Buyback rates or credits would be higher for energy derived from DG located in constrained areas of the distribution system. Finally, DG owners sometimes face the implications of 'stranded costs' of utility investments in restructured markets. Competitive transition charges and exit fees can apply when a DG customer–owner seeks to switch providers or disconnect from the grid entirely.

In the future, one key area of concern is the technical details of interconnecting DG with the electric power systems (EPSs). RES will contribute to meet the targets of the Kyoto Protocol and support the security of supply with respect to limited energy resources. The interconnection must allow DG sources to be connected with the EPS in a manner that provides value to the end user without compromising reliability or performance.

The situation in Europe differs from country to country. Circumstances may also differ between synchronous interconnected systems and island systems. The capacity targets and the future portfolio of RES depend on the national situation. Nevertheless the biggest growth potential is for wind energy. The expectations of the European Wind Energy Association show an increase from 28.5 GW in 2003 to 180 GW in 2020. Due to different support schemes for RES restrictions in licensing and a limited number of suitable locations, this capacity tends to focus on very few regions in Europe. However, new wind farms will normally be built far away from the main load centres. New overhead lines will therefore be necessary to transport the electricity to where it is consumed. These investments are exclusively or at least mainly driven by the new RES generation sites. The intermittent contributions from wind power must be balanced with other backup generation capacity located elsewhere. This adds to the requirements for grid reinforcements.

The licensing procedures for new lines are lasting several years, some even more than 10 years. A delay in grid extension will result in a delay of RES investments because wind farms cannot earn an adequate return on investment without an adequate grid connection. New lines are therefore critical for the success of new RES. Moreover, this new infrastructure could be a significant investment. There is not yet a European-wide harmonized rule about who should pay for it. The legal framework and administrative procedures have to be set properly to speed up the licensing of grid infrastructure.

As countermeasures, suitable European-wide harmonized grid codes for new wind farms and other RES defining their electrical behaviour in critical grid situations are needed in all countries expanding their share of RES. Existing wind farms not fulfilling the actual grid code requirements must be upgraded or replaced (i.e. the electrical behaviour of wind turbines in case of grid faults). Finally, a sufficient capacity from conventional generation has to be in the system at any one time to keep it stable.

1.9   Recommendations and Guidelines for DG Planning

Liberalization and economic efficiency.   Liberalization of the electricity market has increased the complexity and transaction costs for all market players and particularly affected smaller producers. In certain markets where they can avoid charges on transmission, distributed generators may have an advantage over central generators. Elsewhere, in wholesale markets that are designed with large central generation in mind, smaller distributed generators may be at a disadvantage because of the additional costs and complexities of dealing with the market. Difficulties in the New Electricity Trading Arrangements (NETA)/British Electricity Trading and Transmission Arrangements (BETTA) market in the UK suggest that further market measures are needed to make the system fair to smaller generators [25]. Furthermore, treatment of connection charges for DG needs to be consistent with treatment of larger generators. In fact, liberalization of the electricity market is not broad enough to take advantage of the flexibility of many types of DG. Retail pricing should encourage the development of DG in locations where it can reduce network congestion and operate at times when system prices are high.

Environmental protection. DG embraces a wide range of technologies with a wide range of both NO_x and GHG emissions. Emissions per kWh of NO_x from DG (excepting diesel generators) tend to be lower than emissions from a coal-fired power plant. At the same time, the emissions rate from existing DG (excepting fuel cells and photovoltaics) tends to be higher than the best available central generation, such as a combined cycle gas turbine with advanced emissions control. This puts a serious limitation on DG in areas where NO_x emissions are rigorously controlled. If, however, DG is used in a CHP mode, there can be significant emissions savings, even compared with combined cycle power plants. Measures should be designed that encourage distributed generators to reduce their emissions. The use of economic instruments (such as carbon emissions trading) would encourage DG operators to design and operate their facilities in ways that minimize emissions of GHGs.

Regulatory issues and interest in DG. The profits of distribution companies are directly linked to sales. The more kilowatt hours of electricity that move over their lines, the more money they make. Interconnecting with customer-owned DG is plainly not in line with a utility's profit motive. Permission to connect to the grid should be restricted only for safety and grid protection. Guidelines should ensure that there are no restrictions, other than for safety or grid protection reasons.

The following issues need to be addressed [26–31]:

- Adopt uniform technical standards for connecting DG to the grid.
- Adopt testing and certification procedures for interconnection equipment.
- Accelerate development of control technology and systems. While policy increases interest in DG, regulatory and institutional barriers surrounding the effective deployment of DG remain.
- Adopt standard commercial practices for any required utility review of interconnection.
- Establish standard business terms for interconnection agreements.

- Develop tools for utilities to assess the value and impact of distributed power at any point on the grid.
- Develop new regulatory principles compatible with the distributed power choices in competitive and utility markets.
- Adopt regulatory tariffs and utility incentives to fit the new distributed power model. Design tariffs and rates to provide better price transparency to DG.
- Define the condition necessary for a right to interconnect.
- Develop a well-designed policy framework that will reward efficiency and environmental benefits in DG technologies the same way as it does for conventional large-scale generators.
- Include critical strategies for consumer education and cost evaluation tools to deploy DG effectively.
- Design rate for standby charges, interconnection fees, exit fees and grid management charges.

Distributed generators must be allowed to connect to the utility grid. The owners of DG must recognize the legitimate safety and reliability concerns associated with interconnection. Regulators must recognize that the requirements for utility studies and additional isolation equipment will be minimal in the case of smaller DG units.

1.10 Summary

DG has the potential to play a major role as a complement or alternative to the electric power grid under certain conditions. DG is fundamentally distinct from the traditional central plant model for power generation and delivery in that it can deliver energy close to loads within the power distribution network. Three relatively independent sources of pressure, namely restructuring, the need for new capacity and DG technology advancements, are collectively laying the groundwork for the possible widespread introduction of DG. Standards for control/communications should be developed to enable DG better to participate in markets.

But DG is not necessarily a benefit for all players in the electricity sector. Utilities may see customers with on-site generation as problematic because they have different consumption patterns than the average customers. DG usually requires the site to have the same service capacity from the utility as that before installation while the customer is buying less energy, i.e. the load factor on installed utility capacity is reduced.

Depending on the operating scheme and relative performance of the DG system and the power plants supplying the grid, fuel consumption, carbon and other pollutant emissions, and noise pollution can all increase or decrease with DG adoption. For these reasons, DG policy needs to encourage applications that benefit the public, while discouraging those from which the public incurs a net cost. Inherent in this, there is a need to analyse DG costs and benefits and the influence of public policy on DG adoption and operation. While DG may itself become a dominant force in the provision of energy, it is its capability to be used in numerous locations and become integrated into the grid that gives it its greatest value.

Public confidence and interest in DG technologies will depend on the availability of reliable technical support. For example, micro-CHP units and solar water heaters require adjustments to a hot water system at home to deliver carbon savings. DG will only have a chance of success if licensing, regulation and pricing regimes can be made so that users of DG technologies are allowed to benefit equitably. Ultimately, there has to be a clear, consistent and long-term market framework. One can envisage the emergence of networks consisting of a large number of local networks with self-sufficiency (islanding) capabilities, connected by a national grid. Recent technological advances in communications, energy storage and automation can make this possible.

Recent developments in the regulatory arrangements and incentives to connect, particularly renewable technologies, to transmission and distribution networks have meant that the traditional pattern of network usage has altered and this situation is likely to continue. One of the principal changes has been the increase in the volume of DG connected to the network. In certain circumstances, this makes use of the transmission network without being liable for transmission charges, and yet its impact on network flows may lead to additional transmission investment.

References

[1] N. Jenkins, R. Allan, P. Crossley, D. Kirschen and G. Strbac, *Embedded Generation*, The Institution of Electrical Engineers, Stevenage, 2000.

[2] T.J. Hammons and L.L. Lai, 'International practices in distributed generation development worldwide', Paper 07GM0434, IEEE 2007 General Meeting, Tampa, FL, USA, 24–28 June 2007.

[3] D. Pimentel, G. Rodrigues, T. Wane, R. Abrams, K. Goldberg, H. Staecker, E. Ma, L. Brueckner, L. Trovato, C. Chow, U. Govindarajulu and S. Boerke, 'Renewable energy: economic and environmental issues', *BioScience*, Vol. 44, No. 8, 1994.

[4] Rob van Gerwen, *Distribution Generation and Renewables*, Copper Development Association, Hemel Hempstead, November 2006.

[5] CHPQA, 'Defining Good Quality CHP', Guidance Note 10, CHPQA, Department of Environment, Food and Rural Affairs, London, 2000.

[6] Suchismita S. Duttagupta and Chanan Singh, 'A reliability assessment methodology for distribution systems with distributed generation', IEEE 2006 General Meeting, Montreal, Canada, June 2006.

[7] B. Courcelle, 'Distributed generation: from a global market to niche applications', Honeywell, Distributed Power 2001, Nice, France, May 2001.

[8] Kwang Y. Lee, 'The effect of DG using fuel cell under deregulated electricity energy markets', Paper No. 06GM1321, IEEE 2006 General Meeting, Montreal, Canada, June 2006.

[9] Yuping Lu, Xin Yi, Xia Lin and Ji'an Wu, 'An intelligent islanding technique considering load balance for distribution systems with DGs', Paper No. 06GM0323, IEEE 2006 General Meeting, Montreal, Canada, June 2006.

[10] Robert Priddle, *Distributed Generation in Liberalised Electricity Markets*, Report by the International Energy Agency, Paris, 2002.

[11] Peter Fraser, 'The economics of distributed generation', *Energy Prices and Taxes*, 4th Quarter 2002, International Energy Agency, Paris, pp. xi–xviii.

[12] Arthur D. Little, Inc., *Distributed Generation: Understanding the Economics,* Cambridge, MA, 1999.

[13] *Distributed Generation*, A Factfile, The Institution of Engineering and Technology www.theiet.org/factfiles, Stevenage, 2006.

[14] Energy Resources International, Inc., 'Distributed generation in the southern states: barriers to development and potential solutions', for The Southern States Energy Board and The Mississippi Development Authority – Energy Division, April 2003.

[15] B. Alderfer, M. Eldridge and T. Starrs, *Making Connections: Case Studies of Interconnection Barriers and their Impacts on Distributed Power Projects*, National Renewable Energy Laboratory, Golden, CO, 2000.

[16] Lisa Schwartz, *Distributed Generation in Oregon: Overview, Regulatory Barriers and Recommendations*, Public Utility Commission, Oregon, February 2005.

[17] Tze-Fun Chan and Loi Lei Lai, 'Permanent-magnet machines for distributed power generation: a review', Paper No. 07GM0593, IEEE 2007 General Meeting, Tampa, FL, USA, 24–28 June 2007.

[18] http://en.wikipedia.org/wiki/Clean_Development_Mechanism

[19] European Transmission System Operators, 'Integration of renewable energy sources in the electricity system - grid issues', Brussels, 30 March 2005.

[20] *Accommodating Distributed Generation*, Econnect Project No. 1672, Department of Trade and Industry, London, May 2006.

[21] S. Suryanarayanan, W. Ren, M. Steurer, P.F. Ribeiro and G.T. Heydt, 'A Real-time controller concept demonstration for distributed generation interconnection', IEEE 2006 General Meeting, Montreal, Canada, June 2006.

[22] Arthur D. Little, Inc., *Distributed Generation: System Interfaces*, Cambridge, MA, 1999.

[23] Khaled A. Nigim, Ahmed F. Zobaa and Wei Jen Lee, 'Micro grid integration opportunities and challenges', Paper No. 07GM0284, IEEE 2007 General Meeting, Tampa, FL, USA, 24–28 June 2007.

[24] N. Miller and Z. Ye, *Distributed Generation Penetration Study*, National Renewable Energy Laboratory, NREL/SR-560-34715, Golden, CO, August 2003.

[25] *Report to the DTI on the Review of the Initial Impact of NETA on Smaller Generators*, Office of Gas and Electricity Markets, London, August 2001.

[26] *Embedded Generation: Price Controls, Incentives and Connection Charging, a Preliminary Consultation Document*, Office of Gas and Electricity Markets, London, 2001.

[27] Arthur D. Little, Inc., *Distributed Generation: Policy Framework for Regulators*, Cambridge, MA, 1999.

[28] IEA, *Security of Supply in Electricity Markets: Evidence and Policy Issues*, International Energy Agency, Paris, 2002.

[29] *Enduring Transmission Arrangements for Distributed Generation*, Ref. 92/06, Ofgem, London, 31 May 2006.

[30] *Enduring Charging Arrangements for Distributed Generation*, Office of Gas and Electricity Markets (Ofgem), London, September 2005.

[31] Ryan Firestone, Chris Marnay and Karl Magnus Maribu, *The Value of Distributed Generation under Different Tariff Structures*, Report No. LBNL-60589, Ernest Orlando Lawrence Berkeley National Laboratory, May 2006.

2

Generators

2.1 Introduction

Electric power generators are devices that convert mechanical energy into electrical energy. The most common type of electric power generator, such as a bicycle dynamo, uses the principle of electromagnetic induction to convert mechanical energy into electrical energy. These devices carry one or more coils surrounded by a magnetic field, typically supplied by a permanent magnet or electromagnet. In a direct current (DC) generator, a mechanical switch (or commutator) causes the rotor current to reverse every half an electrical cycle so that the output current remains unidirectional. In an alternating current (AC) generator, the rotor is driven by a turbine and electric currents are induced in the stator winding. Large alternators in modern power stations are of this type and they provide the electric power for general transmission and distribution.

Small generators are sometimes movable, whereas large generators are invariably mounted at a fixed location. Energy inputs usually include conventional fuel sources such as diesel and natural gas; however, some electric power generators use alternative forms of energy such as hydro, wave and wind power. In terms of outputs, some electric power generators provide single-phase or three-phase AC voltage. Other generators output DC power.

2.2 Synchronous Generator

Synchronous generators may be one of the following two types, namely, (i) high-speed generators with cylindrical rotors, and (ii) low-speed generators with salient-pole rotors carrying a large number of poles. Type (i) generators are usually large generators with two or sometimes four poles. These are usually driven by steam

Distributed Generation: Induction and Permanent Magnet Generators L. L. Lai and T. F. Chan
© 2007 John Wiley & Sons, Ltd

turbines. Generators of type (ii) may be further divided into two subclasses. The first subclass is very large generators with a large number of poles, e.g. 50. These generators are driven by water turbines and are employed exclusively for power generation. The second subclass consists of smaller independent generators (such as may be used in emergency power supplies), usually with ratings in the range of 10 to 5000 kW, that operate at various submultiples of 3600 r/min.

The stator winding of a synchronous generator carries high currents that are best delivered to the load through fixed terminals. It may also be subjected to large electromagnetic forces during transient operating conditions, therefore the stator end winding must be securely reinforced with mechanical support. Under steady-state conditions, the rotor and stator currents are constant and the rotor moves in synchronism with the rotating flux wave. The rotor flux linkage is constant, hence the induced voltage in the field winding is zero.

2.2.1   Permanent Magnet Materials

Permanent magnet (PM) steel most commonly used in the early part of the last century was mainly an alloy of iron with 1 % carbon. This had an energy product of about 1600 J/m^3. A more expensive steel, with 35 % cobalt and smaller quantities of chromium and tungsten, could have an energy product of about 7500 J/m^3. These steels could be shaped and quenched from a high temperature to give them their hard magnetic and mechanical properties. Permanent magnets could be found in many electrical and electronic products, such as televisions, telephones, computers, audio systems and also in automobiles.

The available product range of permanent magnet materials has increased in recent years. Permanent magnets can in general be classified as non-rare-earth type and rare earth type. Non-rare-earth magnets include Alnico (aluminium–nickel–cobalt) and ceramics (strontium and barium ferrites). Although non-rare-earth magnets are used in the majority of the above applications due to their economic cost, rare earth permanent magnets have many distinguishing characteristics, such as a large maximum energy product. The maximum energy product (which has units of J/m^3) is the maximum value for the product of magnetizing force and induction, and is an indication of the strength of the material. Dozens of rare earth permanent magnet materials have been developed recently, including Sm–Co (samarium–cobalt) magnets and Nd–Fe–B (neodymium–iron–boron) magnets. Nd–Fe–B material has a magnetizing force more than 10 times stronger than a traditional ferrite magnet, and its maximum energy product can easily reach 250 kJ/m^3. References [1] and [2] give information on other permanent magnets. However, the magnetic performance of Nd–Fe–B material will deteriorate rapidly above about 180 °C, and the corrosion and oxidation resistance of Nd–Fe–B is relatively low.

Nd–Fe–B magnets can be made in either sintered or polymer-bonded forms. Polymer-bonded (such as epoxy-bonded) magnets can be produced with close tolerances off tool, with little or no finishing required. Sintered magnets usually require some finishing operations in order to hold close mechanical tolerances, but they possess better magnetic properties than bonded magnets. Sintered Nd–Fe–B permanent magnets are manufactured by using a powder metallurgical process. At first, an Nd–Fe–B alloy is formulated according to the desired magnetic properties and produced in a vacuum furnace. The resultant product is crushed into powder form and then pressed. During the pressing process, magnetic fields are applied with the assistance of a specially designed fixture to align the magnetic 'domains' and to optimize the magnetic performance. The pressed magnets are then placed in a furnace under a protective atmosphere for sintering. After sintering, the magnet shape is rough and needs to be machined and ground to achieve the desired shape and size. A surface coating is usually applied on Nd–Fe–B magnets. Zinc or nickel coating is commonly used but other materials such as cadmium chromate, aluminium chromate, tin or polymer (epoxy) can also be used. Each permanent magnet material has its own pros and cons, so how to choose the right one for a particular application is a challenge to any user. A balance between cost and performance must be considered when selecting a permanent magnet material.

2.2.2   Permanent Magnet Generator

The use of permanent magnet field systems in electric machines has found increasing acceptance in recent years. This is partly due to the need for inexpensive and reliable excitation systems, for which ferrite magnets are particularly well suited. A more important development, however, is that the application of new permanent magnet materials (such as the rare earths) in new configurations has resulted in a high specific output or in other characteristics that are difficult to achieve with non-permanent-magnet machines. Permanent magnet materials are now available with a wide range of characteristics, allowing considerable scope in the choice of magnet composition [3].

The elimination of the risk of demagnetization under fault conditions and the need for a high power/volume ratio has encouraged the use of the permanent magnet generator (PMG), which is a synchronous machine in which the rotor windings have been replaced with permanent magnets. Compared with a wound-field machine, a PMG offers a higher output for a given frame size and it can be operated over a wide speed range. The rotor excitation loss, which otherwise typically represents 20 to 30 % of the total generator losses in a wound-field machine, is eliminated. The reduced losses result in a lower temperature rise in the generator, which means that a smaller and simpler cooling system can be used. Generator reliability is improved by a prolonged life span of the bearings as a result of reduced operating temperature. Perhaps the most important feature of synchronous generators is that they can operate at or close to unity power factor. Summarizing, the advantages of a PMG include brushless construction, light weight, small size, high reliability, less frequent maintenance and high efficiency. The disadvantage, however, is that the excitation cannot be varied and hence the output voltage of the generator will vary with load. From practical considerations, it is desirable that the voltage regulation of the generator be minimized. This may be accomplished by capacitor compensation, an electronic voltage controller, or using a generator with inherent voltage regulation capability.

Radial flux PMG for isolated operation. Conventional PMGs are generally of the radial flux type. The rotor configuration may be surface magnet type, interior type or surface inset type [4]. PMGs developed in the late 1970s and early 1980s employed low-cost ceramic magnets. Binns and Kurdali [5] developed a PMG of a novel multi-stacked form. The use of capacitance to improve voltage regulation and to increase power output was discussed. In another paper [6], Binns and Low described the performance and application of multi-stacked imbricated PMGs. Typical characteristics were discussed and it was shown that relatively cheap anisotropic ferrite magnets are well suited to this type of machine. For windmill applications, switched load with a fixed capacitor in parallel was used for approximate load matching between the turbine and generator power characteristics. Voltage control of permanent magnet synchronous generators (PMSGs) by using shunt capacitors was also studied by Rahman *et al.* [7] and Chen *et al.* [8].

Chen *et al.* [9] studied a radial flux PMSG with outer-rotor construction that facilitated direct coupling to the wind turbine. The initial electromagnetic design was based on classical magnetic circuit analysis but the finite element method (FEM) was used to obtain the detailed characteristics.

Chalmers [10] subsequently showed that a PMG with interior magnets has an inverse saliency feature (i.e. the d-axis synchronous reactance X_d is less than the q-axis synchronous reactance X_q). Negative voltage regulation could result when the PMG supplied an isolated resistive load. Towards the mid-1980s, Ne–Fe–B emerged as an important class of high-energy PM material [4]. Since then it has been widely used in electric machines, such as brushless DC motors and PMGs.

Some countries in the world have abundant natural resources that could be exploited for distributed power generation. The western regions of China, for example, have natural gas reserves but are too remote for central grid access. A DG system based on small-gas-turbine technology is therefore an option worthy of consideration. By using a direct-driven generator operating at very high speeds (typically above 30 000 r/min), there is a significant reduction in system size and an improvement in efficiency. The main technical issues [10, 11] to be addressed are the electromagnetic designs of the PMG for high-speed operation (such as the rotor structure and the choice of pole number), reduction of iron and stray losses, and development of high-speed bearings. Figure 2.1 shows the possible rotor construction for a high-speed PMG, where the magnets are secured by a retaining ring or layers of fibre-glass bands against the centrifugal forces due to rotation.

Linear PM Machines. Linear electric machines are electromagnetic devices that involve translational motion. In the past linear induction motors (LIMs) and linear synchronous motors (LSMs) have been developed mainly for high-speed ground

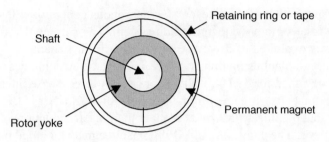

Figure 2.1 *Four-pole rotor for high-speed PMG. Reproduced by permission of the IEEE (© 2007 IEEE)*

transportation systems, conveyors, people movers, and propulsion systems for air-craft take-off. For generator operation, the motion of the linear generator has to be of short distance and oscillatory in nature. A brief review of the history, types and applications of linear electric actuators and generators was given by Boldea and Nasar [12]. Figure 2.2 shows the construction of a simple moving PM linear generator.

Amara *et al.* [13] investigated a tubular linear PM machine that might offer the highest efficiency and power/force density. The machine studied had a nine-slot, ten-pole configuration, with a fractional number of slots per pole per phase. Since the Halbach magnetic array was used for the mover, the flux at the inner bore is quite small, permitting the use of a non-magnetic supporting tube for the magnets. The problem of eddy current reduction is the main design issue for this type of machine configuration.

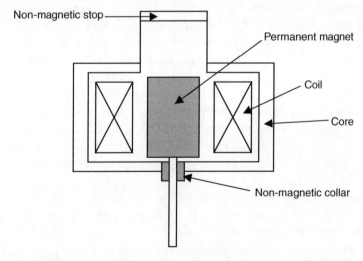

Figure 2.2 *A simple linear oscillatory PMG. Reproduced by permission of the IEEE (© 2007 IEEE)*

A typical application of the linear PM synchronous generator (LPMSG) is in a wave energy to electric energy conversion system. The perpetual vertical motion of sea waves is exploited to drive the mover of a linear generator for producing electricity. A pilot plant using this principle, namely the Archimedes Wave Swing (AWS), was designed for 4 MW peak power [14]. A three-phase linear PMG was designed to extract electric energy from the motion of the floater [15]. Polinder *et al.* [16] showed that the linear PMG was cheaper than linear induction generators for such applications. The conventional LPMSG had magnets mounted on the mover, but the authors also proposed a transverse flux PMG (TFPMG) which had flux concentrators, magnets and conductors all on the stator, while the translator only consisted of iron. Such a machine configuration was more difficult to build and hence further investigations would be needed.

Axial flux PM machines. Axial flux machines are usually disk shaped with flat, annulus air gaps. The mutual flux is oriented in the axial direction while the effective conductors of the armature run in radial directions. Axial flux PMGs have recently received considerable attention, mainly in wind energy conversion systems (WECs) as they could be designed for use with direct-coupled wind turbines. Since the bulky and expensive mechanical gearbox is eliminated, the noise level is reduced and the reliability of supply is improved. The axial flux PMSG can be conveniently designed with a high pole number that is required for low-speed operation.

A modular PMG with axial flux direction was proposed for a direct-drive wind turbine generator by Muljadi *et al.* [17]. Flexibility of design was achieved since each generator could be expanded to several modules. Each unit module in turn was built from simple modular poles. The stator winding was formed like a torus, linked by flux-focusing guides. The prototype built could deliver a power of 650 W at an efficiency of 75 % when driven at 667 r/min.

Chalmers and colleagues [18–20] developed an axial flux PMG called 'torus' for a gearless wind energy system. The stator armature consisted of a laminated toroidal core on which the air gap winding was wound, as in the classical Gramme-ring winding. Cogging torque was therefore eliminated by the slotless design, while the double-sided PM rotor minimizes the magnetic pull between the stator and rotor. The experimental machine in [19] had a rating of 5 kW at 200 r/min. Output voltage was practically sinusoidal and an efficiency of 82 % could be accomplished at full load. When the air gap lengths on the two sides of the torus are not the same, the resultant magnet pull in the axial direction can be significant, and special thrust bearings need to be used.

Hwang *et al.* [21] and Parvianen *et al.* [22] adopted the more conventional axial flux machine configuration with a toothed iron core. Hwang's machine [21] employed the doubled-sided configuration in which the rotor PM disk was sandwiched between two toothed stator armature cores. The 24-pole machine was rated at 10 kVA and 380 V when operating at 300 r/min. Parvianen *et al.* [22] on the other hand developed a single-sided axial machine configuration with open slots. Various combinations of armature slot number and pole number were studied with a view to reducing the cogging torque. A 1.6 kW prototype machine was constructed and was installed in a pilot power plant. Since a single-sided configuration was used, thrust bearings had to be employed to withstand the strong magnetic pull between the stator and rotor.

Spooner *et al.* [23] proposed an ironless, radial flux, PM for a direct-coupled wind generator machine that employed lightweight spoke-wheel structures for both the rotor and the stator. A working flux density of about 0.25 T was produced at the winding. The generator could have a mass typically 20–30 % of equivalent designs based on iron-cored magnetic circuits, and the efficiency was greater than 90 %. The experimental machine had a specification of 11.1 kW at 150 r/min.

Chan and Lai [24] recently reported a single-sided axial flux PMSG for small-scale wind energy systems with either vertical axis or horizontal axis turbines. The machine featured an outer-rotor configuration and had an air gap disk armature winding, hence there was no cogging torque and no iron loss. At a speed of 600 r/min the generator could deliver a power of 350 W at rated current. Figure 2.3 shows the test rig for load tests on the prototype axial flux PMG.

Summarizing, various designs of axial flux PMGs could be adopted for direct-coupled wind turbine applications. The choice of a particular design depends upon the expertise of individual research teams, the availability of components and materials (e.g. Nd–Fe–B magnets, wound silicon cores, etc.) and the accessible manufacturing technology.

Axial flux PMGs could also be used for high-speed applications. For example, Wang *et al.* [25] reported on the optimal design of a coreless, double-sided stator axial flux PMSG suitable for integration with internal combustion engines (ICEs), e.g. as integrated starter–generators for use in hybrid vehicles.

Figure 2.3 *Test rig for experimental investigations on a prototype axial flux PMSG. Reproduced by permission of the IEEE (© 2007 IEEE)*

Variable speed PMG connected to grid. Variable speed power generation enables the turbine generator system to operate at maximum power conditions. Since the frequency of a PMG varies with the speed, a frequency converter has to be provided for integration with the power grid. Kimura *et al.* [26] presented a distributed power generation system that included a variable speed turbine with a power conversion system for use in a cogeneration system. The PMG was to be directly driven by a diesel engine operating at variable speed, while the output power was delivered to a distribution line. Experimental results were presented and some practical operational issues were discussed. Amei *et al.* [27] proposed a maximum power control strategy for a wind-driven PMSG by using a rectifier, a boost chopper circuit and an inverter, and a theoretical analysis was carried out. Chinchilla *et al.* [28] also investigated maximum power control of a direct-coupled wind turbine generator by using a pulse-width-modulated (PWM) rectifier, an intermediate DC circuit and a PWM inverter. In addition, vector control of the grid-side inverter allows power factor regulation of the system. The dynamic system performance was analysed and experimentally verified.

2.3   Induction Generator

Over the past few decades, there has been an increasing use of squirrel-cage-type induction generators (IGs), particularly in wind energy systems and micro-hydro power systems. The grid provides frequency and voltage regulation, as well as the reactive power required by the IG. Due to the distributed nature of the energy resources, these power systems are usually small scale in terms of rating. They may not be as efficient as central bulk power systems, but this disadvantage is offset by the reduction or even elimination of the transmission losses over long distances. The global trend of privatization and deregulation is a further impetus to the development of small-scale distributed or (embedded) generation systems [29, 30].

Even in developed countries, energy conservation and environmental protection can be achieved by extensive renewable energy programs and more widespread use of waste heat and cogeneration [31]. For such applications, the low cost and flexibility of using IGs result in their increasing popularity.

In remote regions of some developing countries, rural electrification is often based on single-phase generation and transmission/distribution systems [32]. This approach has the advantage that, for a given amount of capital investment, a wider area can be provided with electricity. There is thus a great need for the development of single-phase IGs. Although single-phase induction motors may be adapted for generator operation, it is often more economical, for ratings above 3 kW, to use standard three-phase induction machines [33]. With a suitable phase balancing scheme, explained in a later chapter, the three-phase IG can operate satisfactorily on a single-phase grid. A practical phase balancing scheme invariably employs passive circuit elements, such as capacitance, inductance or resistance.

In countries and regions where a grid connection is difficult and expensive to provide, a more cost-effective solution is to develop stand-alone or autonomous power systems [34, 35] which may consist of one or more induction generators and a small number of loads to be served, comprising typically lighting, heating and water pumping. The absence of the grid implies that the reactive power has to be furnished locally by capacitors, and an induction machine operating in this mode is often referred to as a self-excited induction generator (SEIG). Both the output voltage and frequency depend on the connected load and will vary greatly when the load or the speed of the prime-mover is changed. Frequency control and voltage control are two important operational aspects for autonomous power systems. Both three-phase systems and single-phase systems may be developed. Besides, the choice of appropriate values of excitation capacitances to secure successful voltage build-up, and to sustain the voltage when the generator is supplying load, is of paramount importance.

Single-phase operation of a three-phase IG, with or without a phase balancer, means that the machine operates with phase imbalance. For the grid-connected IG, the voltage and frequency are constant, hence the analysis can be carried out by using the method of symmetrical components [36]. With the SEIG, however, the performance analysis is complicated by the variation of frequency as well as the magnetizing reactance, both being dependent on the speed and loading conditions.

2.3.1   Three-Phase IGs and SEIGs

The principle and operation of grid-connected three-phase IGs are well understood and are discussed in detail in many textbooks [37]. Performance analysis is based on the induction motor equivalent circuit, negative values of slip being used since the rotor speed is higher than the synchronous speed. The SEIG, on the other hand, involves more complicated analysis and has received considerable attention. Pioneering work on the SEIG dates back to the 1920s and 1930s [38, 39] when it was discovered that an induction machine with capacitance connected to the stator terminals might stay excited after being disconnected from the grid. The significance of such a phenomenon for generator application was apparent, but since then very little has been written on the subject and the synchronous generator dominates the role for large-scale power generation.

Towards the late 1970s and early 1980s, however, interest in SEIGs revived as witnessed by the numerous research publications. Murthy *et al.* [40] and Malik and Hague [41] analysed the SEIG using the loop impedance method, based on the per-phase equivalent circuit model. The Newton–Raphson method was employed for simultaneously determining the per-unit frequency and magnetizing reactance. Ouazene and McPherson [42], on the other hand, proposed the nodal admittance method, also based on the equivalent circuit. This approach resulted in a high-order polynomial in the per-unit frequency which could be solved to yield the generator performance. The generalized induction machine model has been used by Elder *et al.* [43], Grantham *et al.* [44] and Wang [45] for the analysis of voltage build-up and transient operation of the SEIG.

The capacitor sizing problem for a three-phase SEIG was studied by Malik and Mazi [46], Jabri and Alolah [47] and Chan [48], while the performance of an SEIG driven by regulated and unregulated turbines was investigated by Bonert and Hoops [49], Chan [50] and Alghuwainem [51]. Wind-turbine-driven SEIGs were studied by Ammasaigounden *et al.* [52], Watson *et al.* [53] and Raina and Malik [54].

Voltage compensation using the long-shunt connection was first investigated by Bim *et al.* [55] with a view to improving the voltage regulation characteristic of the SEIG. This method, together with the short-shunt configuration, was subsequently analysed by Chan [56] and Wang and Su [57]. Application of the compensated SEIG for supplying an induction motor load has also been reported [58].

Voltage and frequency control of the three-phase SEIG has been studied fairly recently [59–64]. Voltage control invariably involves the regulation of effective capacitance across the stator terminals, and a variety of control methods, such as the fixed-capacitor thyristor-controlled reactor (FC-TCR) [41], might be employed. Frequency control for the SEIG, on the other hand, is more difficult and involves expensive and sophisticated equipment, such as an AC/DC/AC converter [59, 64].

2.3.2   Single-Phase IGs and SEIGs

The analysis and performance of a grid-connected single-phase IG with main and auxiliary windings were reported by Boardman *et al.* [65]. It was found that the generator efficiency was higher if the rotor was driven to give reverse rotation (i.e. the rotor rotates against the air gap travelling field). This principle was also applicable to a three-phase IG connected to a single-phase power grid [66]. When a three-phase IG operates on a single-phase power system, the currents are generally unbalanced and the rated three-phase power output cannot be developed. Other adverse effects include thermal overload, mechanical vibration, noise, poor efficiency and low power factor. Various phase balancing schemes for a three-phase IG have been investigated [33, 67, 68]. In the method proposed by Durham and Ramakumar [33], an artificial third line is created, using a capacitance and an inductance of equal reactance, for supplying the IG whose power factor has been corrected to unity. The disadvantage of such a scheme is that, in the event that the supply is removed, severe overvoltages will be produced as a result of series resonance between the capacitance and inductance. In the phase balancing schemes proposed by Smith [67], capacitors were used exclusively and there was no danger of a resonance effect. The analysis, however, was confined to the case when the induction machine was exactly balanced. Most of the circuits introduced require a ground wire in the supply system, which may not be feasible in some regions. More recently, Chan [68] investigated phase balancing for an IG using the Steinmetz connection, and it was demonstrated that perfect phase balance could be achieved over a wider range of generator impedance angle by the use of dual-phase converters.

The study on single-phase SEIGs was conducted by Murthy [69], Murthy *et al.* [70], Rahim *et al.* [71], Chan [72] and Singh and Shilpkar [73]. The two-phase symmetrical component method was applied to an SEIG with main and auxiliary windings in quadrature. Ojo [74] presented a transient analysis of single-phase SEIGs using the *d–q* equivalent circuit model. Application of single-phase SEIG for heating and lighting loads in remote regions was proposed by Singh *et al.* [75] and microprocessor control of single-phase SEIG was reported by Watson and Watson [76]. Ojo *et al.* [77] also investigated the operation of a single-phase SEIG using a PWM inverter with a battery supply.

2.4   Doubly Fed Induction Generator

Doubly fed induction generator (DFIG) wind turbines are nowadays increasingly used in large wind farms [78, 79]. The main reason for the popularity of wind-driven DFIGs connected to the national grid is their ability to supply power at constant voltage and frequency as the rotor speed varies. The DFIG configuration also provides a possibility to control the overall system power factor. One form of DFIG uses a wound-rotor induction machine in which the rotor is fed via a frequency converter to give variable speed operation.

A typical DFIG system is shown in Figure 2.4. The AC/DC/AC converter consists of two components: the rotor-side converter (R_{con}) and the grid-side converter (G_{con}). R_{con} and G_{con} are voltage-sourced converters that use forced–commutated power electronics devices (IGBTs) to synthesize an AC voltage from a DC voltage source. A capacitor connected on the DC side acts as the DC voltage source. The generator slip rings are connected to the rotor-side converter, which shares a DC link with the grid-side converter in a so-called back-to-back configuration. The AC side of the grid-side converter is connected through a three-phase series inductance to the power network, while the stator winding of the generator is directly connected to the network. The power captured by the wind turbine is converted into electric power by the IG and is transmitted to the grid by the stator and the rotor windings. The control system generates the pitch angle command and the voltage command signals for R_{con} and G_{con} to control the power of the wind turbine, DC bus voltage and reactive power or voltage at the grid terminals.

2.4.1   Operation

When the rotor moves faster than the rotating magnetic field from the stator, it means that the stator induces a strong current in the rotor. The harder the rotor rotates, the more power will be transferred as an electromagnetic force to the stator, and in turn converted to electricity which is fed into the electrical grid. The speed of the asynchronous generator will vary with the turning force applied to it. Its difference from the synchronous speed in per cent is called the generator's

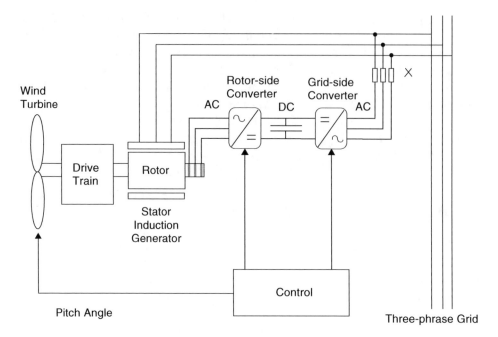

Figure 2.4   *A DFIG and wind turbine system*

slip. With the rotor winding short-circuited, the generator at full load is only a few per cent.

With the DFIG, slip control is provided by the rotor- and grid-side converters. At high rotor speeds, the slip power is recovered and delivered to the grid, resulting in a high overall system efficiency. If the rotor speed range is limited, the ratings of the frequency converters will be small compared with the generator rating, which helps in reducing converter losses and the system cost.

Since the mechanical torque applied to the rotor is positive for power generation and since the rotational speed of the magnetic flux in the air gap of the generator is positive and constant for a constant frequency grid voltage, the sign of the rotor electric power output is a function of the slip sign. The rotor electric power output is positive for negative slip (speed greater than synchronous speed) and negative for positive slip (speed lower than synchronous speed). For supersynchronous speed operation, the rotor electric power output is transmitted to the DC bus capacitor and tends to raise the DC voltage. For subsynchronous speed operation, the rotor electric power output is taken out of the DC bus capacitor and tends to decrease the DC voltage. G_{con} is used to keep the DC voltage constant. At steady state for a lossless AC/DC/AC converter, the grid-side converter electric power output is equal to the rotor electric power output and the speed of the wind turbine is determined by the rotor electric power output absorbed or generated by R_{con}. R_{con} and G_{con} have the capability of generating or absorbing reactive power and could be used to control the reactive power or the voltage at the grid terminals. The pitch angle is controlled to limit the generator output power to its nominal value for high wind speeds. The reactive power absorbed by the induction generator is provided by the grid or by some devices such as capacitor banks or synchronous condensers.

2.4.2   Recent Work

Most wind turbines run at almost constant speed with direct grid connection. With indirect grid connection, the wind turbine generator can operate independently of the grid so that the frequency of the stator current may be varied. This permits the turbine to run at variable rotational speed for maximum power tracking while delivering AC to the grid at the system frequency. The conversion from variable frequency AC to DC can be carried out with converters that employ thyristors or large IGBTs, while the fluctuating DC is converted to AC by using a line-commutated inverter. The output alternating voltage and current are not pure sinusoids. The current waveform can be smoothed out by using appropriate inductors and capacitors, but the non-smooth appearance of the voltage waveform does not disappear completely. There are also additional energy losses in the AC–DC–AC conversion process.

The transient performance of a DFIG under network disturbances has been studied. DFIG models, which include the main performance characteristics of the generator, have been developed for transient stability studies [80–83]. Dynamic modelling of DFIGs for variable speed wind energy generation plants has also been reported in [84–87]. One common approach in dynamic modelling of DFIGs for wind turbines is to use a model based on the space vector theory of a slip-ring induction machine [88]. This method provides sufficient accuracy also in cases where the voltage dips due to single- or two-phase faults in the network. More advanced transient analysis that takes into account magnetic saturation and the skin effect in the rotor winding would require another model of the DFIG, e.g. the FEM model. Comparisons of stabilizing methods for DFIGs and stability under faults have been discussed in [81–83].

The impact on power quality was studied in [89]. During a fault, the voltage drop will lead to shaft and gearbox mechanical stresses and overcurrent surges. The rotor-side converter could have difficulties in handling these current spikes. To protect the converter and to assist the wind turbine to support the grid, various control strategies have been considered and reported in [90–94]. DFIGs are increasingly used in wind turbines built at power levels above 1.5 MW for offshore applications. Speed control is necessary in order to limit the mechanical stresses and power surge for these high-power systems [95–97].

Zhang and colleagues [98, 99] reported an application of a matrix converter for the power control of a variable speed wind turbine driving a DFIG. The control is based on a stator flux vector control algorithm [98], the main objective being to limit the high current in the rotor during grid faults. A variable speed, constant frequency DFIG system using a back-to-back converter for the wind turbine has been developed. Based on an analysis of the stator-flux-oriented vector control theory of DFIGs, a vector control scheme has been developed for power generation [100].

A DFIG for use with a variable speed wind turbine and supplying an isolated load was designed using two back-to-back PWM voltage-fed inverters [101]. DFIGs with power converters are used to develop a slip power recovery scheme that improves efficiency and reduces rotor loss of DFIGs [102, 103]. Apparently various types of converters have been used in DFIG systems [95, 98].

DFIGs are an economic variable speed solution for large wind turbines, while high-voltage DC (HVDC) transmission has been used for grid connection in some offshore wind farms [104–106].

2.5   Summary

PM synchronous machines are suitable for distributed power generation applications. Advances in PM material technology have stimulated the development of more efficient and more compact generator units. To have a significant impact on practical applications, further work on PM machines is needed and may include, but is not necessarily limited to, the following: refined performance analysis, loss and thermal models, innovative machine and winding configurations, design optimization, analytic methods and FEM for field computations, control for isolated operation and grid connection, and other niche application areas.

Asynchronous (or induction) generators can also be used to generate AC. The IG is very reliable, and tends to be comparatively inexpensive. The generator also has some electrical properties which are useful for wind power applications, e.g. a certain overload capability. This type of generator is widely used in wind energy conversion systems, as well as in some small hydro power plants. The use of IGs in grid or isolated power systems has been discussed.

Various aspects regarding the DFIG, such as its operation, control, performance, impact on power quality and wind energy production, have been briefly reviewed.

A comprehensive list of references is included to assist readers in further study.

References

[1] Lester R. Moskowitz, *Permanent Magnet Design and Application Handbook*, Krieger, Malabar, FL, 1995.

[2] http://www.stanfordmagnets.com/magnet.html#nfb

[3] K.J. Binns and T.M. Wong, 'Analysis and performance of a high-field PM synchronous machine', *IEE Proceedings*, Pt. B, Vol. 131, No. 6, pp. 252–258, November 1984.

[4] J.F. Gieras and M. Wing, *Permanent Magnet Motor Technology – Design and Applications*, New York: Marcel Dekker, 1997.

[5] K.J. Binns and A. Kurdali, 'Permanent-magnet a.c. generators', *IEE Proceedings – Electric Power Applications,* Vol. 129, No. 7, pp. 690–696, July 1979.

[6] K.J. Binns and T.S. Low, 'Performance and application of multi-stacked imbricated permanent-magnet generators', *IEE Proceedings,* Pt B, Vol. 130, pp. 407–414, November 1983.

[7] M.A. Rahman, A.M. Osheiba, T.S. Radwan and E.S. Abdin, 'Modelling and controller design of an isolated diesel engine PM synchronous generator', *IEEE Transactions on Energy Conversion,* Vol. 11, No. 2, pp. 324–330, June 1996.

[8] Z. Chen, E. Spooner, W.T. Norris and A.C. Williamson, 'Capacitor-assisted excitation of permanent-magnet generators', *IEE Proceedings – Electric Power Applications,* Vol. 145, No. 6, pp. 497–507, November 1998.

[9] Jian Yi Chen, Chern Nayar and Longya Xu, 'Design and FE analysis of an outer-rotor PM generator for directly-coupled wind turbine applications', *Proceedings of the IEEE–IAS 33rd Annual Meeting,* Vol. 1, pp. 387–394, 1998.

[10] B.J. Chalmers, 'Performance of interior type permanent-magnet alternator', *IEE Proceedings – Electric Power Applications,* Vol. 141, No. 4, pp. 186–190, July 1994.

[11] Fengxiang Wang, Wenpeng Zheng, Ming Zhong and Baoguo Wang, 'Design considerations of high-speed PM generators for micro turbines', *Proceedings of the International Conference on Power System Technology 2002,* Vol. 1, pp. 158–162, 13–17 October 2002.

[12] I. Boldea and S.A. Nasar, 'Linear electric actuators and generators', *IEEE Transactions on Energy Conversion,* Vol. 14, No. 3, pp. 712–717, September 1999.

[13] Yacine Amara, Jiabin Wang and David Howe, 'Analytical prediction of eddy-current loss in modular tubular permanent-magnet machines', *IEEE Transactions on Energy Conversion,* Vol. 20, No. 4, pp. 761–770, December 2005.

[14] Mats Leijon, Hans Bernhoff, Olov Agren, Jan Isberg, Jan Sundberg, Marcus Berg, Karl Erik Karlson and Arne Wolfbrandt, 'Multiphysics simulation of wave energy to electric energy conversion by PM linear generator', *IEEE Transactions on Energy Conversion,* Vol. 20, No. 1, pp. 219–224, March 2005.

[15] H. Polinder, M.E.C. Damen and F. Garder, 'Linear PM generator for wave energy conversion in the AWS', *IEEE Transactions on Energy Conversion,* Vol. 19, No. 3, pp. 583–589, September 2004.

[16] H. Polinder, Barrie C. Mecrow, Alan G. Jack, Phililip G. Dickinson and Markus A. Mueller, 'Conventional and TFPM linear generators for direct-drive wave energy conversion', *IEEE Transactions on Energy Conversion,* Vol. 20, No. 2, pp. 260–267, June 2005.

[17] E. Muljadi, C.P. Butterfield and U.H. Wan, 'axial flux modular permanent-magnet generator with a toroidal winding for wind-turbine applications', *IEEE Transactions on Industry Application,* Vol. 35, No. 4, pp. 831–836, July/August 1999.

[18] W. Wu, E. Spooner and B.J. Chalmers, 'Design of slotless TORUS generators with reduced voltage regulation', *IEE Proceedings on Electric Power Applications,* Vol. 142, No. 5, pp. 337–343, September 1995.

[19] B.J. Chalmers, W. Wu and E. Spooner, 'An axial flux PM generator for gearless wind energy system', *IEEE Transactions on Energy Conversion,* Vol. 14, No. 2, pp. 251–257, June 1999.

[20] J.R. Bumby, R. Martin, M.A. Mueller, E. Spooner, N.L. Brown and B.J. Chalmers, 'Electromagnetic design of axial flux permanent magnet machines', *IEE Proceedings – Electric Power Applications,* Vol. 151, No. 2, pp. 151–159, March 2004.

[21] D. Hwang, K. Lee, D. Kang, Y. Kim, K. Choi and D. Park, 'An modular-type axial flux PM synchronous generator for gearless wind power systems', *Proceedings of the IEEE–IES 30th Annual Conference (IECON 2004),* Pusan, Korea, Vol. 2, pp. 1396–1399, 2–6 November 2004.

[22] A. Parvianen, J. Pyrhonen and P. Kontkanen, 'Axial flux permanent magnet generator with concentrated winding for small wind power applications', *Proceedings of the 2005 IEEE*

International Conference on Electric Machines and Drives, San Antonio, Texas, USA, pp. 1187–1191, May 2005.

[23] E. Spooner, P. Gordon, J.R. Bumby and C.D. French, 'Lightweight ironless-stator PM generators for direct-drive wind turbines', *IEE Proceedings on Electric Power Applications*, Vol. 152, No. 1, pp. 17–26, January 2005.

[24] T.F. Chan and L.L. Lai, 'An axial-flux permanent-magnet synchronous generator for a direct-coupled wind turbine system', *IEEE Transactions on Energy Conversion*, Vol. 22, No. 1, pp. 86–94, March 2007.

[25] Rong-jie Wang, Maarten J. Kamper, Kobus Van der Westhuizen and Jacek F. Gieras, 'Optimal design of a coreless stator axial flux permanent-magnet generator', *IEEE Transactions on Energy Conversion*, Vol. 41, No. 1, pp. 55–63, January 2005.

[26] M. Kimura, H. Koharagi, K. Imaie, S. Dodo, H. Arita and K. Tsubouchi, 'A permanent-magnet synchronous generator with variable-speed input for co-generation system', *Proceedings of the IEEE–PES Winter Meeting 2001*, Vol. 3, pp. 1419–1424, 26 January–1 February 2001.

[27] Kenji Amei, Yuichi Takayasu, Takahisa Ohji and Massaki Sakui, 'A maximum power control of wind generator system using a permanent magnet synchronous generator and a boost chopper circuit', *Proceedings of the Power Conversion Conference 2002*, Osaka, Japan, Vol. 3, pp. 1447–1452, 2–5 April 2002.

[28] Monica Chinchilla, Santiago Arnaltes and Juan Carlos Burgos, 'Control of permanent-magnet generators applied to Variable speed wind-energy systems connected to the grid', *IEEE Transactions on Energy Conversion*, Vol. 21, No. 1, pp. 130–135, March 2006.

[29] Loi Lei Lai (ed.), *Power System Restructuring and Deregulation: Trading, Performance and Information Technology*, John Wiley & Sons, Ltd, Chichester, 2001.

[30] N. Jenkins, 'Embedded generation', *IEE Power Engineering Journal*, Vol. 9, No. 3, pp. 145–150, June 1995.

[31] R.L. Nailen, 'Watts from waste heat - induction generators for the process industries', *IEEE Transactions on Industry Applications*, Vol. IA-19, No. 3, pp. 470–475, May/June 1983.

[32] P.G. Holmes, 'Single- to 3-phase transient phase conversion in induction motor drives', *IEE Proceedings*, Pt B, Vol. 132, No. 5, pp. 289–296, September 1985.

[33] M.O. Durham and R. Ramakumar, 'Power system balancers for an induction generator', *IEEE Transactions on Industry Applications*, Vol. IA-23, No. 6, pp. 1067–1072, November/December 1987.

[34] R. Holland, 'Appropriate technology - rural electrification in developing countries', *IEE Review*, Vol. 35, No. 7, pp. 251–254, August 1989.

[35] A. Doig, 'Off-grid electricity for developing countries', *IEE Review*, Vol. 45, No. 1, pp. 25–28, January 1999.

[36] J.E. Brown and O.I. Butler, 'A general method of analysis of 3-phase induction motors with asymmetrical primary connections', *IEE Proceedings*, Vol. 100, Pt II, pp. 25–34, February 1953.

[37] M.G. Say, *Alternating Current Machines*, 5th edn, Pitman (ELBS) London, pp. 333–336, 1983.

[38] E.D. Bassett and F.M. Potter, 'Capacitance excitation for induction generators', *Transactions of AIEE*, Vol. 54, pp. 540–545, May 1935.

[39] C.F. Wagner, 'Self-excitation of induction motors with series capacitors', *Transactions of AIEE*, Vol. 60, pp. 1241–1247, 1941.

[40] S.S. Murthy, O.P. Malik and A.K. Tandon, 'Analysis of self-excited induction generators', *IEE Proceedings*, Pt C, Vol. 129, No. 6, pp. 260–265, November 1982.

[41] N.H. Malik and S.E. Hague, 'Steady state analysis and performance of an isolated self-excited induction generator', *IEEE Transactions on Energy Conversion*, Vol. EC-1, No. 3, pp. 134–139, September 1986.

[42] L. Ouazene and G. McPherson, Jr, 'Analysis of the isolated induction generator', *IEEE Transactions on Power Apparatus and Systems*, Vol. PAS-102, No. 8, pp. 2793–2798, August 1983.

[43] J.M. Elder, J.T. Boys and J.L. Woodward, 'Self-excited induction machine as a small low-cost generator', *IEE Proceedings*, Pt C, Vol. 131, No. 2, pp. 33–41, March 1984.

[44] C. Grantham, D. Sutanto and B. Mismail, 'Steady-state and transient analysis of self-excited induction generators', *IEE Proceedings*, Pt B, Vol. 136, No. 2, pp. 61–68, March 1989.

[45] L. Wang, 'A novel analysis on the performance of an isolated self-excited induction generator', *IEEE Transactions on Energy Conversion*, Vol. 12, No. 2, pp. 109–115, June 1997.

[46] N.H. Malik and A.A. Mazi, 'Capacitance requirements for isolated self excited induction generators', *IEEE Transactions on Energy Conversion*, Vol. EC-2, No. 1, pp. 62–68, March 1987.

[47] A.K. Jabri and A.I. Alolah, 'Capacitance requirement for isolated self-excited induction generator', *IEE Proceedings*, Pt B, Vol. 137, No. 3, pp. 154–159, May 1990.

[48] T.F. Chan, 'Capacitance requirements of self-excited induction generator', *IEEE Transactions on Energy Conversion*, Vol. 8, No. 2, pp. 304–311, June 1993.

[49] R. Bonert and G. Hoops, 'Standalone induction generator with terminal impedance controller and no turbine control', *IEEE Transactions on Energy Conversion*, Vol. EC-5, No. 1, pp. 28–31, March 1990.

[50] T.F. Chan, 'Self-excited induction generators driven by regulated and unregulated turbines', *IEEE Transactions on Energy Conversion*, Vol. 11, No. 2, pp. 338–343, June 1996.

[51] S.M. Alghuwainem, 'Steady-state analysis of an isolated self-excited induction generator driven by regulated and unregulated turbines', *IEEE Transactions on Energy Conversion*, Vol. 14, No. 2, pp. 718–723, June 1999.

[52] N. Ammasaigounden, M. Subbiah and M.R. Krishnamurthy, 'Wind-driven self-excited pole-changing induction generators', *IEE Proceedings*, Pt B, Vol. 133, No. 5, pp. 315–321, September 1986.

[53] D.B. Watson, J. Arrillaga and T. Densem, 'Controllable d.c. power supply from wind driven self-excited induction machines', *IEE Proceedings*, Vol. 126, No. 12, pp. 1245–1248, December 1979.

[54] G. Raina and O.P. Malik, 'Wind energy conversion using a self-excited induction generator', *IEEE Transactions on Power Apparatus and Systems*, Vol. PAS-102, No. 12, pp. 3933–3936, December 1983.

[55] E. Bim, J. Szajner and Y. Burian, 'Voltage compensation of an induction generator with long-shunt connection', *IEEE Transactions on Energy Conversion*, Vol. EC-4, No. 3, pp. 526–530, September 1989.

[56] T.F. Chan, 'Analysis of self-excited induction generators using an iterative method', *IEEE Transactions on Energy Conversion*, Vol. 10, No. 3, pp. 502–507, September 1995.

[57] L. Wang and J.Y. Su, 'Effects of long-shunt and short-shunt connections on voltage variations of a self-excited induction generator', *IEEE Transactions on Energy Conversion*, Vol. 12, No. 4, pp. 368–374, December 1997.

[58] L. Wang and C.H. Lee, 'Long-shunt and short-shunt connections on dynamic performance of a SEIG feeding an induction motor load', *IEEE Transactions on Energy Conversion*, Vol. 15, No. 1 pp. 1–7, March 2000.

[59] E. Profumo, B. Colombo and F. Mocci, 'A frequency controller for induction generators in stand-by minihydro power plants', *Proceedings of the 4th International Conference on Electrical Machines and Drives*, IEE Conference Publication No. 310, 13–15 September 1989, London, UK.

[60] M.A. Al-Saffa, E.-C. Nho, and T.A. Lipo, 'Controlled shunt capacitor self-excited induction generator', *Thirty-Third IEEE Industry Applications Society Annual Meeting Conference Record*, Vol. 2, pp. 1486–1490, 1998.

[61] R. Bonert and S. Rajakaruna, 'Self-excited induction generator with excellent voltage and frequency control', *IEE Proceedings – Generation, Transmission and Distribution*, Vol. 145, No. 1, pp. 33–39, January 1998.

[62] O. Chtchetinine, 'Voltage stabilization system for induction generator in stand alone mode', *IEEE Transactions on Energy Conversion*, Vol. 14, No. 3, pp. 298–303, September 1999.

[63] E. Suarez and G. Bortolotto, 'Voltage-frequency control of a self-excited induction generator', *IEEE Transactions on Energy Conversion*, Vol. 14, No. 3, pp. 394–401, September 1999.

[64] E.G. Marra and J.A. Pomilio, 'Induction-generator-based system providing regulated voltage with constant frequency', *IEEE Transactions on Industrial Electronics*, Vol. 47, No. 4, pp. 908–914, August 2000.

[65] E.C. Boardman, S.S. Venkata and N.G. Butler, 'The effect of rotational direction in single-phase induction generators', *IEEE Transactions on Power Apparatus and Systems*, Vol. PAS-103, No. 8, pp. 2222–2229, August 1984.

[66] G.L. Johnson, *Wind Energy Systems*, Prentice Hall, Englewood Cliffs, NJ, 1985.

[67] O.J.M. Smith, 'Three-phase induction generator for single-phase line', *IEEE Transactions on Energy Conversion*, Vol. EC-2, No. 3, pp. 382–387, September 1987.

[68] T.F. Chan, 'Performance analysis of a three-phase induction generator connected to a single-phase power system', *IEEE Transactions on Energy Conversion*, Vol. 13, No. 3, pp. 205–211, September 1998.

[69] S.S. Murthy, 'A novel self-excited self-regulated single phase induction generator Part 1: basic system and theory', *IEEE Transactions on Energy Conversion*, Vol. 8, No. 3, pp. 377–382, September 1993.

[70] S.S. Murthy, H.C. Rai and A.K. Tandon, 'A novel self-excited self-regulated single phase induction generator Part 2: experimental investigation', *IEEE Transactions on Energy Conversion*, Vol. 8, No. 3, pp. 383–388, September 1993.

[71] Y.H.A. Rahim, A.I. Alolah and R.I. Al-Mudaiheem, 'Performance of single phase induction generators', *IEEE Transactions on Energy Conversion*, Vol. 8, No. 3, pp. 389–395, September 1993.

[72] T.F. Chan, 'Analysis of a single-phase self-excited induction generator', *Electric Machines and Power Systems*, Vol. 23, No. 2, pp. 149–162, March 1995.

[73] B. Singh and L.B. Shilpkar, 'Steady-state analysis of the single-phase self-excited induction generator', *IEE Proceedings – Generation, Transmission and Distribution*, Vol. 146, No. 5, pp. 421–427, September 1999.

[74] O. Ojo, 'The transient and qualitative performance of a self-excited single-phase induction generator', *IEEE Transactions on Energy Conversion*, Vol. 10, No. 3, pp. 493–501, September 1995.

[75] B. Singh, R.B. Saxena, S.S. Murthy and B.P. Singh, 'A single-phase self-excited induction generator for lighting loads in remote areas', *International Journal of Electrical Engineering Education*, Vol. 25, No. 3, pp. 269–275, July 1988.

[76] D.B. Watson and R.M. Watson, 'Microprocessor control of a self-excited induction generator', *International Journal of Electrical Engineering Education*, Vol. 22, No. 1, pp. 83–92, January 1985.

[77] O. Ojo, O. Omozusi, A. Ginart and G. Gonoh, 'The operation of a stand-alone, single-phase induction generator using a single-phase, pulse-width modulated inverter with a battery supply', *IEEE Transactions on Energy Conversion*, Vol. 14, No. 3, pp. 526–531, September 1999.

[78] http://www.mathworks.com/access/helpdesk/help/toolbox/physmod/powersys

[79] S. Muller, M. Deicke and R.W. De Doncker, 'Doubly fed induction generator systems for wind turbines', *Industry Applications Magazine, IEEE*, Vol. 8, No. 3, pp. 26–33, May–June 2002.

[80] O. Anaya-Lara, F.M. Hughes, N. Jenkins and G. Strbac, 'Power system stabiliser for a generic DFIG-based wind turbine controller', *Proceedings of the 8th International Conference on AC and DC Power Transmission, IEE*, pp. 145–149, 2006.

[81] L. Holdsworth, X.G. Wu, J.B. Ekanayake, and N. Jenkins, 'Comparison of fixed speed and doubly-fed induction wind turbines during power system disturbances', *IEE Proceedings – Generation, Transmission and Distribution*, Vol. 150, No. 3, pp. 343–352, May 2003.

[82] P. Ledesma and J. Usaola, 'Doubly fed induction generator model for transient stability analysis', *IEEE Transactions on Energy Conversion*, Vol. 20, No. 2, pp. 388–397, June 2005.

[83] M.V.A. Nunes, J.A.P. Lopes, H.H. Zurn, U.H. Bezerra and R.G. Almeida, 'Influence of the Variable speed wind generators in transient stability margin of the conventional generators integrated in electrical grids', *IEEE Transactions on Energy Conversion*, Vol. 19, No. 4, pp. 692–701, December 2004.

[84] O. Anaya-Lara, F.M. Hughes, N. Jenkins and G. Strbac, 'Contribution of DFIG-based wind farms to power system short-term frequency regulation', *IEE Proceedings – Generation, Transmission and Distribution*, Vol. 153, No. 2, pp. 164–170, March 2006.

[85] J.B. Ekanayake, L. Holdsworth, X.G. Wu and N. Jenkins, 'Dynamic modeling of doubly fed induction generator wind turbines', *IEEE Transactions on Power Systems*, Vol. 18, No. 2, pp. 803–809, May 2003.

[86] L. Holdsworth, X.G. Wu, J.B. Ekanayake and N. Jenkins, 'Direct solution method for initialising doubly-fed induction wind turbines in power system dynamic models', *IEE Proceedings – Generation, Transmission and Distribution*, Vol. 150, No. 3, pp. 334–342, May 2003.

[87] J. Ekanayake and N. Jenkins, 'Comparison of the response of doubly fed and fixed-speed induction generator wind turbines to changes in network frequency', *IEEE Transactions on Energy Conversion*, Vol. 19, No. 4, pp. 800–802, December 2004.

[88] A. Tapia, G. Tapia, J.X. Ostolaza and J.R. Saenz, 'Modeling and control of a wind turbine driven doubly fed induction generator', *IEEE Transactions on Energy Conversion*, Vol. 18, No. 2, pp. 194–204, June 2003.

[89] Y. Liao, L. Ran, G.A. Putrus and K.S. Smith, 'Evaluation of the effects of rotor harmonics in a doubly-fed induction generator with harmonic induced speed ripple', *IEEE Transactions on Energy Conversion*, Vol. 18, No. 4, pp. 508–515, December 2003.

[90] J. Morren and S.W.H. de Haan, 'Ride through of wind turbines with doubly-fed induction generator during a voltage dip', *IEEE Transactions on Energy Conversion*, Vol. 20, No. 2, pp. 435–441, June 2005.

[91] B. Xie, B. Fox and D. Flynn, 'Study of fault ride-through for DFIG based wind turbines', *Proceedings of the International Conference on Electric Utility Deregulation, Restructuring and Power Technologies, DRPT2004, IEEE*, Vol. 1, pp. 411–416, April 2004.

[92] R. Cardenas, R. Pena, J. Proboste, G. Asher and J. Clare, 'MRAS observer for sensorless control of standalone doubly fed induction generators', *IEEE Transactions on Energy Conversion*, Vol. 20, No. 4, pp. 710–718, December 2005.

[93] H. Banakar, C. Luo and B.T. Ooi, 'Steady-state stability analysis of doubly-fed induction generators under decoupled P-Q control', *IEE Proceedings – Electric Power Applications*, Vol. 153, No. 2, pp. 300–306, March 2006.

[94] F.M. Hughes, O. Anaya-Lara, N. Jenkins and G. Strbac, 'Control of DFIG-based wind generation for power network support', *IEEE Transactions on Power Systems*, Vol. 20, No. 4, pp. 1958–1966, November 2005.

[95] M. Bartram, J. von Bloh and R.W. De Doncker, 'Doubly-fed-machines in wind-turbine systems: is this application limiting the lifetime of IGBT-frequency-converters?' *Proceedings of the 35th Annual Power Electronics Specialists Conference PESC04*, Vol. 4, pp. 2583–2587, 2004.

[96] K.S. Rongve, B.I. Naess, T.M. Undeland and T. Gjengedal, 'Overview of torque control of a doubly fed induction generator', *Proceedings of the Power Tech Conference, IEEE*, Vol. 3, June 2003.

[97] S. Muller, M. Deicke, and R.W. De Doncker, 'Adjustable speed generators for wind turbines based on doubly-fed induction machines and 4-quadrant IGBT converters linked to the rotor', *Proceedings of the Industry Applications Conference, IEEE*, Vol. 4, pp. 2249–2254, October 2000.

[98] L. Zhang, C. Watthanasarn and W. Shepherd, 'Application of a matrix converter for the power control of a Variable speed wind-turbine driving a doubly-fed induction generator', *Proceedings of the 23rd International Conference on Industrial Electronics, Control and Instrumentation, IECON97*, Vol. 2, pp. 906–911, November 1997.

[99] L. Zhang and C. Watthanasarn, 'A matrix converter excited doubly-fed induction machine as a wind power generator', *Proceedings of the Seventh International Conference on Power Electronics and Variable Speed Drives*, IEE Conference Publlication No. 456, pp. 532–537, September 1998.

[100] R. Pena, J.C. Clare and G.M. Asher, 'Doubly fed induction generator using back-to-back PWM converters and its application to variable-speed wind-energy generation', *IEE Proceedings – Electric Power Applications*, Vol. 143, No. 3, pp. 231–241, May 1996.

[101] R. Pena, J.C. Clare and G.M. Asher, 'A doubly fed induction generator using back-to-back PWM converters supplying an isolated load from a variable speed wind turbine', *IEE Proceedings – Electric Power Applications*, Vol. 143, No. 5, pp. 380–387, September 1996.

[102] B. Rabelo and W. Hofmann, 'Optimal active and reactive power control with the doubly-fed induction generator in the MW-class wind-turbines', *Proceedings of the 4th International Conference on Power Electronics and Drive Systems*, Vol. 1, pp. 53–58, October 2001.

[103] A. Muthuramalingam and V.V. Sastry, 'The DC link series resonant converter based slip power recovery scheme - an implementation', *Proceedings of the International Conference on Power Electronics and Drive Systems, PEDS99*, Vol. 1, pp. 121–126, July 1999.

[104] D. Xiang, L. Ran, J.R. Bumby, P.J. Tavner and S. Yang, 'Coordinated control of an HVDC link and doubly fed induction generators in a large offshore wind farm', *IEEE Transactions on Power Delivery*, Vol. 21, No. 1, pp. 463–471, January 2006.

[105] A. Petersson, T. Thiringer, L. Harnefors and T. Petru, 'Modeling and experimental verification of grid interaction of a DFIG wind turbine', *IEEE Transactions on Energy Conversion*, Vol. 20, No. 4, pp. 878–886, December 2005.

[106] Y. Lei. A. Mullane, G. Lightbody and R. Yacamini, 'Modeling of the wind turbine with a doubly fed induction generator for grid integration studies', *IEEE Transactions on Energy Conversion*, Vol. 21, No. 1, pp. 257–264, March 2006.

3

Three-Phase IG Operating on a Single-Phase Power System

3.1 Introduction

In this chapter, the general principle of phase balancing for a three-phase IG operating on a single-phase power system is investigated and several practical phase balancing schemes are proposed, including those that involve dissipative elements and current injection transformers. It is demonstrated that the IG–converter system can be analysed by using the method of symmetrical components. A phasor diagram approach enables the conditions of perfect phase balance to be deduced. Performance analysis of a three-phase IG with the Smith connection is also possible using the same approach. The feasibility of the phase balancing schemes is verified by laboratory experiments on a small induction machine. A microcontroller-based control scheme for an IG with the Smith connection is also proposed to give efficient control at low cost.

3.2 Phase Balancing using Passive Circuit Elements

3.2.1 Analysis of IG with Phase Converters

Plain single-phase operation of a three-phase machine is an extreme case of unbalanced operation. This stems from the fact that the line current flowing into the 'free' terminal of the stator winding is forced to be zero. To reduce the phase imbalance, an effective remedy is to inject a line current artificially into the 'free' terminal by using phase converters which comprise passive circuit elements. Figure 3.1 illustrates the principle of phase balancing for an induction machine operating on a single-phase power system [1–3]. The rotor is assumed to be rotating in such a

Distributed Generation: Induction and Permanent Magnet Generators L. L. Lai and T. F. Chan
© 2007 John Wiley & Sons, Ltd

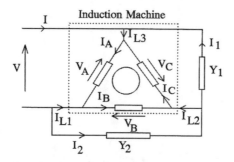

Figure 3.1   *Single-phase operation of three-phase IG with phase converters*

direction that it traverses the stator winding in the sequence A–B–C. For generator operation, the rotor speed must be slightly higher than the positive-sequence rotating field. Although special reference is made to a delta-connected machine in the following discussion, the principle is also applicable to a star-connected machine. A-phase of the IG is connected to the single-phase power system of voltage V, while the phase converters Y_1 and Y_2 are respectively connected across C-phase and B-phase. The current I_{L2} that results from the currents I_1 and I_2 through the phase converters constitutes the line current into the 'free' terminal of the generator. Apparently the phase balance is improved and indeed, by appropriate choice of the values of Y_1 and Y_2, perfect phase balance may be achieved.

Referring to Figure 3.1 and adopting the *motor* convention for the induction machine, the following 'inspection equations' [4] may be written:

$$V = V_A \tag{3.1}$$

$$V_A + V_B + V_C = 0 \tag{3.2}$$

$$I_1 = V_C Y_1 \tag{3.3}$$

$$I_2 = V_B Y_2 \tag{3.4}$$

$$I_1 = I_B - I_C + I_2. \tag{3.5}$$

Solving the above equations in association with the symmetrical component equations given in Appendix A, the positive-sequence voltage V_p and negative-sequence voltage V_n can be determined as follows:

$$V_p = \sqrt{3}V \cdot \frac{Y_n + \frac{e^{-j\pi/6}}{\sqrt{3}}Y_1 + \frac{e^{j\pi/6}}{\sqrt{3}}Y_2}{Y_1 + Y_2 + Y_p + Y_n} \tag{3.6}$$

$$V_n = \sqrt{3}V \cdot \frac{Y_p + \frac{e^{j\pi/6}}{\sqrt{3}}Y_1 + \frac{e^{-j\pi/6}}{\sqrt{3}}Y_2}{Y_1 + Y_2 + Y_p + Y_n} \tag{3.7}$$

where Y_p and Y_n are, respectively, the positive-sequence and negative-sequence admittances of the IG as shown in Appendix A.

For perfect phase balance, the negative-sequence voltage component V_n given by (3.7) should be equal to zero, hence

$$Y_p + \frac{e^{j\pi/6}}{\sqrt{3}} Y_1 + \frac{e^{-j\pi/6}}{\sqrt{3}} Y_2 = 0. \tag{3.8}$$

By selecting values of Y_1 and Y_2 that satisfy (3.8), balanced operation of the IG may be achieved.

3.2.2 Phase Balancing Schemes

Based on the theory outlined in Section 3.2.1, four practical phase balancing schemes for a three-phase IG operating on a single-phase power system have been developed and investigated. Figures 3.2(a)–(d) show the details of the circuit connections. For convenience of discussion, each phase balancing scheme is designated by the phase converter elements used, suffix 1 denoting a C-phase converter element and suffix 2 a B-phase converter element. For example, an R_1–C_2 scheme will have a resistance connected across C-phase and a capacitance connected across B-phase.

The $C_1(L_1)$–C_2 scheme shown in Figure 3.2(a) employs only energy storage elements, while the remaining schemes employ dissipative (or lossy) elements in addition to energy storage elements. For IG applications, the phase converter resistances can take the form of storage heating elements, auxiliary loads or battery

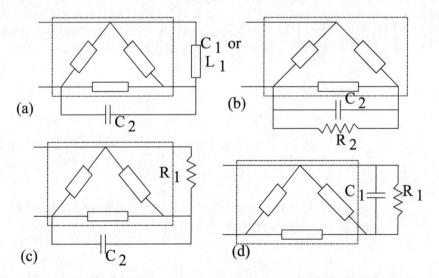

Figure 3.2 *Phase balancing schemes for three-phase IG: (a) $C_1(L_1)$–C_2 scheme; (b) R_2–C_2 scheme; (c) R_1–C_2 scheme; (d) R_1–C_1 scheme*

chargers. From the IG system's point of view, the power dissipated in these loads may be regarded as useful output as far as efficiency evaluation is concerned.

Using (3.8), it is possible to determine the values of the phase converters that result in perfect phase balance. As an illustration, consider the $C_1(L_1)$–C_2 scheme shown in Figure 3.2(a). Assuming Y_1 and Y_2 to be pure *capacitances* (i.e. $Y_1 = 0 + jB_1$; $Y_2 = 0 + jB_2$), (3.8) may be written as

$$Y_p e^{-j\phi_p} + \frac{1}{\sqrt{3}} B_1 e^{j2\pi/3} + \frac{1}{\sqrt{3}} B_2 e^{j\pi/3} = 0 \tag{3.9}$$

where Y_p is the positive-sequence admittance of the generator and ϕ_p is the positive-sequence impedance angle.

Equating real and imaginary parts respectively to zero in (3.9), the values of the phase converter susceptances B_1 and B_2 are given by

$$B_1 = \sqrt{3}G_p + B_p = 2Y_p \sin(2\pi/3 - \phi_p) \tag{3.10}$$

$$B_2 = -\sqrt{3}G_p + B_p = 2Y_p \sin(\phi_p - \pi/3). \tag{3.11}$$

When $\phi_p = 2\pi/3$ rad, $B_1 = 0$ and $B_2 = \sqrt{3}Y_p$, Figure 3.2(a) is reduced to the well-known Steinmetz connection [5].

Table 3.1 summarizes the values of phase converter elements to give perfect phase balance for various phase balancing schemes. It is observed that the values of the phase converter elements are functions of the conductances and susceptances (or alternatively, the admittance and impedance angle) of the positive-sequence IG equivalent circuit.

To check the feasibility of the above phase balancing schemes, experiments were performed on a 2.2 kW, 220 V, 50 Hz, four-pole, delta-connected induction machine

Table 3.1   *Conductances and susceptances of phase converters for perfect phase balance of three-phase IG*

Scheme	Conductance (S)	Susceptance[a] (S)
$C_1(L_1)$–C_2	$G_1 = 0$	$B_1 = 2Y_p\sin(2\pi/3 - \phi_p)$
	$G_2 = 0$	$B_2 = 2Y_p\sin(\phi_p - \pi/3)$
R_2–C_2	$G_1 = 0$	$B_1 = 0$
	$G_2 = \sqrt{3}Y_p\sin(\phi_p - 2\pi/3)$	$B_2 = \sqrt{3}Y_p\cos(\phi_p - 2\pi/3)$
R_1–C_2	$G_1 = 2\sqrt{3}Y_p\sin(\phi_p - 2\pi/3)$	$B_1 = 0$
	$G_2 = 0$	$B_2 = 2\sqrt{3}Y_p\sin(5\pi/6 - \phi_p)$
R_1–C_1	$G_1 = \sqrt{3}{}_p\cos(5/6 - \phi_p)$	$B_1 = \sqrt{3}Y_p\sin(5\pi/6 - \phi_p)$
	$G_2 = 0$	$B_2 = 0$

[a] Capacitive susceptances defined to be positive; inductive susceptances defined to be negative.

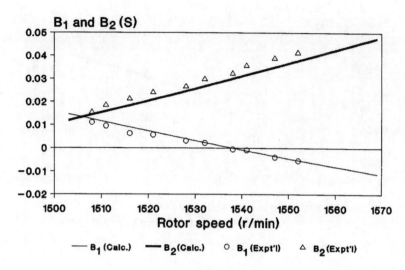

Figure 3.3   *Values of B_1 and B_2 to give phase balance at different speeds in $C_1(L_1)$–C_2 scheme*

IG1whose parameters are given in Appendix D.1. It was found that exact phase balance could in general be obtained with appropriate choice of values of phase converters, subject to the limitations inherent in each phase balancing scheme. Figures 3.3 to 3.6 show the variation of the phase converter conductances/susceptances that result in perfect phase balance when the IG is operating on a 220 V single-phase power system. Very good agreement between the computed and experimental results is observed, thus verifying the theory developed in Section 3.2.1.

3.2.3   Case Study

The performance, limitation and application of each phase balancing scheme are discussed as follows.

$C_1(L_1)$–C_2 *scheme.*   As shown in Figure 3.3, perfect phase balance can be achieved over the practical operating speed range of the IG (1500 to 1570 r/min). The susceptance B_2 increases approximately linearly with speed and remains capacitive over the whole speed range. On the other hand, B_1 decreases with speed. At speeds below 1539 r/min B_1 is capacitive and above this speed B_1 is inductive. At 1539 r/min, phase balance can be achieved with a single capacitance across B-phase.

R_2–C_2 *scheme.*   Using this scheme, perfect phase balance is possible only when the positive-sequence impedance angle ϕ_p exceeds $2\pi/3$ rad (which results in positive values of the phase converter conductance G_2). Both G_2 and B_2 increase with increase in rotor speed (Figure 3.4). This scheme is useful when a large percentage of the input power needs to be delivered to the power system.

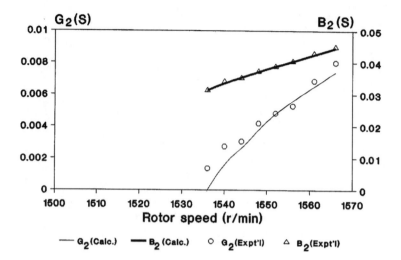

Figure 3.4   *Values of G_2 and B_2 to give phase balance at different speeds in R_2–C_2 scheme*

R_1–C_2 *scheme.* As in the previous scheme, perfect phase balance is possible only when ϕ_p exceeds $2\pi/3$ rad. The conductance G_1 increases with speed, but the susceptance B_2 remains substantially constant (Figure 3.5).

R_1–C_1 *scheme.* As shown in Figure 3.6, perfect phase balance over the normal generator speed range is possible with this scheme. Since G_1 has a much larger value compared with conductances in the previous two schemes, a larger amount of power is dissipated in the phase converter. This scheme is useful when a large percentage of the prime-mover power is to be consumed by the local loads.

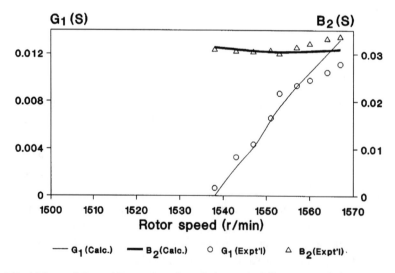

Figure 3.5   *Values of G_1 and B_2 to give phase balance at different speeds for R_1–C_2 scheme*

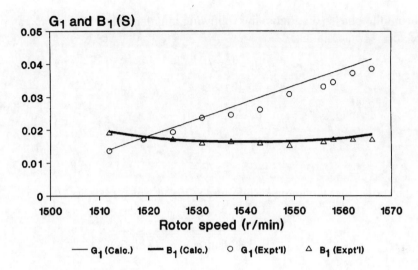

Figure 3.6 Values of G_1 and B_1 to give phase balance at different speeds for R_1–C_1 scheme

3.2.4 System Power Factor

The output power factor of the IG system can in general be computed using the equations presented in Section 3.2.1. For perfect phase balance, however, a closed-form expression for the system power factor can be deduced from the voltage–current relationship in the phasor diagram. As an illustration, Figure 3.7(a) shows the phasor diagram for the $C_1(L_1)$–C_2 scheme, assuming that the phase voltages and currents are balanced and ϕ_p is less than $2\pi/3$ rad, while Figure 3.7(b) shows the relationship between the system input current I and the generator line current

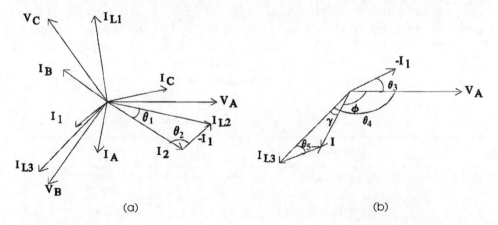

Figure 3.7 Phasor diagrams for $C_1(L_1)$–C_2 scheme under perfect phase balance: (a) phasor diagram under perfect phase balance; (b) phasor diagram showing detailed angular relationships

I_{L1}. From the phasor diagrams, the following angular relationships can easily be deduced:

$$\theta_1 = 2\pi/3 - \phi_p$$

$$\theta_2 = 2\pi/3$$

$$\theta_3 = \pi/6$$

$$\theta_4 = \phi_p + \pi/6$$

$$\theta_5 = 2\pi/3 - \phi_p.$$

It can be shown that the angle γ in Figure 3.7(b) is given by

$$\sin \gamma = \frac{2\sin^2\alpha}{\sqrt{3 + 4\sin^2\alpha - 2\sqrt{3}\sin 2\alpha}} \qquad (3.12)$$

where

$$\alpha = 2\pi/3 - \phi_p. \qquad (3.13)$$

Hence the power factor angle ϕ is

$$\phi = \theta_4 - \gamma = \phi_p + \pi/6 - \gamma. \qquad (3.14)$$

The power factor under balanced condition thus depends only on the positive-sequence impedance angle of the IG. The system power factors of the phase balancing schemes are summarized in Table 3.2.

Figure 3.8 shows the variation of system power factor with speed for various phase balancing schemes. It is observed that above 1539 r/min, the $C_1(L_1)$–C_2 and R_2–C_2 schemes have practically the same power factor, which is well over 0.9. The power factor of the R_1–C_2 scheme is slightly lower but is quite acceptable. The power factor of the R_1–C_1 scheme, however, is very low and becomes zero at a rotor speed of 1543 r/min. Below this speed, power is drawn from the power system in order to furnish the power to the phase converter resistance. A line-side power factor correction capacitor may thus be necessary for this phase balancing scheme.

Table 3.2 *System power factors of phase balancing schemes under perfect phase balance*

Scheme	System power factor (p.u.)
$C_1(L_1)$–C_2	$\cos(\phi_p + \pi/6 - \gamma)$ where $\gamma = \sin^{-1}[2\sin^2\alpha/(3 + 4\sin^2\alpha - 2\sqrt{3}\sin 2\alpha)]$. (for $\phi_p < 2\pi/3$, $\alpha = 2\pi/3 - \phi_p$; for $\phi_p > 2\pi/3$, $\alpha = \phi_p - 2\pi/3$)
R_2–C_2	$\cos(\phi_p + \pi/6)$
R_1–C_2	$\cos(3\pi/2 - \phi_p)$
R_1–C_1	$\cos(\phi_p - \pi/6)$

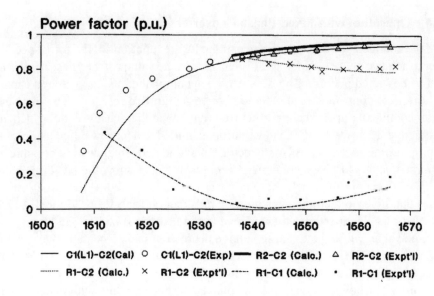

Figure 3.8   *System power factor under perfect phase balance*

3.2.5   Power and Efficiency

With perfect phase balance achieved by the use of phase converters, the IG operates as if it were operating under a balanced three-phase voltage supply. If the power dissipated in the phase converter resistance is regarded as useful power output, the system efficiency under single-phase operation is identical to the generator efficiency under balanced three-phase operation. For the experimental machine, a full-load efficiency of 0.82 was obtained with various phase balancing schemes.

Table 3.3 shows the experimental values of output power to grid, power dissipated in the phase converters, and developed electric power for different phase balancing schemes at a rotor speed of 1552 r/min and a stator current of 5.0 A. In all cases the developed power approaches the rated power of the induction machine. In the R_1–C_1 scheme, however, 86 % of the net output electric power is dissipated in the phase converter resistance, which is consistent with the observation made in Section 3.2.4.

Table 3.3   *Power to grid, power dissipated in phase converter and net electrical output*

Scheme	Power to grid (W)	Power dissipated in phase converter (W)	Net electric power output (W)
$C_1(L_1)$–C_2	1910	0	1910
R_2–C_2	1740	233	1973
R_1–C_2	1564	420	1984
R_1–C_1	270	1606	1876

3.2.6 Operation with Fixed Phase Converters

Both the conductance and susceptance of the IG are functions of the rotor speed. In order to achieve phase balance at different speeds, the values of the phase converters need to be varied accordingly. This can be accomplished using switched capacitors, thyristor-controlled reactors and chopper-controlled resistors. The increased circuit complexity and extra cost incurred may make the phase balancing scheme unattractive. If, however, the load variation is limited to a narrow range using some form of turbine speed control, satisfactory machine performance may be achieved with fixed values of phase converters. Figures 3.9–3.11 show the performance of the R_2–C_2 scheme in which the phase converters are fixed at the experimental values that give perfect phase balance at full-load current. From Figure 3.9, it is observed that the B-phase voltage increases when the speed is decreased from the rated value, while the C-phase voltage decreases. The percentage overvoltage, however, is relatively small. At a speed of 1520 r/min, the experimental B-phase voltage is only 7 % above the rated value.

Figure 3.10 shows the variation of phase and line currents with speed. Again a slight overcurrent occurs in B-phase as the speed is decreased from the rated value. Nevertheless, the increase in copper loss in B-phase is more than offset by the reduction in copper losses in the other two phases, implying that the thermal performance of the generator is satisfactory.

Figure 3.11 shows the variation in power factor and efficiency with speed. As the rotor speed is decreased from the rated value, the IG becomes overcompensated, causing a rise in the system power factor. But as the speed further decreases, the

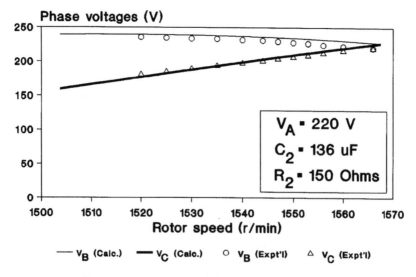

Figure 3.9 *Phase voltages for R_2–C_2 scheme with fixed values of phase converters*

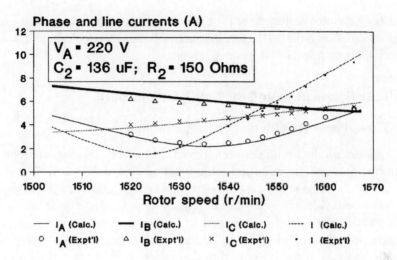

Figure 3.10 *Phase and line currents for $R_2–C_2$ scheme with fixed values of phase converters*

power delivered to the power system is smaller and the power factor drops rapidly. An efficiency close to 0.8 is obtained at speeds above 1550 r/min.

3.2.7 Summary

The feasibility of phase balancing for a three-phase IG operating on a single-phase power system has been investigated. A general analysis for the IG with phase converters is presented, and expressions for the determination of the phase converter elements are given. The effects of phase balancing on the output power, system power factor and efficiency are discussed. It is also demonstrated that

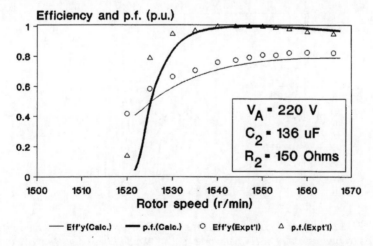

Figure 3.11 *Efficiency and power factor for $R_2–C_2$ scheme with fixed values of phase converters*

satisfactory generator performance is obtained with fixed phase converter elements if the speed variation is limited by turbine control. The theoretical analysis is verified by experiments on a small induction machine.

3.3   Phase Balancing using the Smith Connection

3.3.1   Three-Phase IG with the Smith Connection

Published work on the Smith connection to date has been mainly concerned with the motoring mode of operation and operational aspects such as selection of capacitances for perfect balance and motor starting. Chan and Lai [6, 7] have proposed a method of analysis for these modes of operation.

The objective of this section is to investigate the performance of a three-phase IG with the Smith connection when operating on a single-phase power grid. A systematic analysis of the IG with Smith connection (SMIG) will be presented, and the conditions for perfect phase balance will be deduced. It will be shown that, for medium and heavy loads, perfect phase balance in the induction machine is possible by using only capacitive phase converters. With dual-mode control, satisfactory operation over the normal speed range can be obtained. Experimental results will be used to validate the theoretical analysis.

Figure 3.12 shows the Smith connection for a three-phase IG operating on a single-phase power system. The 'starts' of the stator phases A, B and C are denoted by 1, 2, 3 and the 'finishes' are denoted by 4, 5, 6. Terminals 4 and 6 are common and form the 'pseudo-neutral' point N. Terminals 1 and 2 are both connected to one line of the single-phase grid, while terminal 3 is connected to the second line. Terminal 5 (the 'free' terminal) is connected to the neutral point N via capacitance

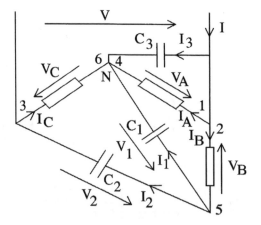

Figure 3.12 *Smith connection for three-phase IG operating on a single-phase grid. Reproduced by permission of T.-F. Chan and L. L. Lai, 'Single-phase operation of a three-phase induction generator with the Smith connection,' IEEE Transactions on Energy Conversion, **17**: 2002, 47–54. © (2002) IEEE*

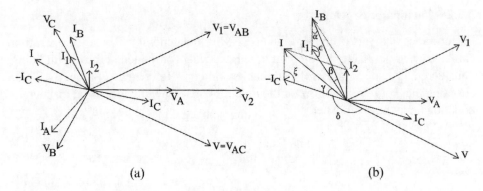

(a) (b)

Figure 3.13 *SMIG under perfect phase balance: (a) phasor diagram; (b) phasor diagram showing angular relationship between currents. Reproduced by permission of T.-F. Chan and L. L. Lai, 'Single-phase operation of a three-phase induction generator with the Smith connection,' IEEE Transactions on Energy Conversion, 17: 2002, 47–54. © (2002) IEEE*

C_1 and to terminal 3 via capacitance C_2. It should be noted that the generator performance is sensitive to the phase connection for a given direction of rotor rotation. A comparison of Figure 3.12 with the Smith connection for the motoring mode [6] reveals that phases B and C have been interchanged, a condition necessary for proper phase balancing as will be explained in the following section.

The Smith connection is essentially an asymmetrical winding connection, but, with an appropriate choice of the terminal capacitances, it is possible for the IG to operate with balanced phase currents and phase voltages. As shown in Figure 3.12, the B-phase current is the sum of the capacitor currents I_1 and I_2. Consider the phasor diagram in Figure 3.13(a), drawn for the special case for perfect phase balance. The current I_1 leads V_1 (or V_{AB}) by $\pi/2$ rad and hence lags V_B by $2\pi/3$ rad. The voltage V_2 (which is equal to $V_{AB} - V_C$) is equal to $2V_A$. The capacitor current I_2 leads V_2 by $\pi/2$ rad and hence it lags V_B by $5\pi/6$ rad. For generator impedance angles between $2\pi/3$ and $5\pi/6$ rad, the phase current I_B can be synthesized with the required magnitude and phase angle to give phase balance, by using suitable values of C_1 and C_2.

Under perfect phase balance conditions, the phase currents of the IG must sum to zero. This requires the currents I_2 and I_3 to be equal, implying that C_3 must be equal to twice of C_2.

With balanced currents flowing in the stator phases, a perfect rotating magnetic field is produced. The air gap voltages per phase, and hence the phase voltages, will also be balanced. The generator operates as if it were supplying a balanced three-phase load, hence the efficiency is the same as that obtaining when the generator operates on a balanced three-phase grid.

Balanced conditions are valid for a given set of capacitance values and speed only. When the rotor speed changes, the circuit conditions are disturbed and a new set of capacitance values is required to balance the generator again.

3.3.2   Performance Analysis

Performance analysis of the three-phase SMIG for general off-balance operation can be carried out using the method of symmetrical components, the circuit being considered as a special case of winding asymmetry. Referring to Figure 3.12 and adopting the *motor* convention for the induction machine, the following inspection equations may be established:

$$V = V_A - V_C \tag{3.15}$$

$$I_B = I_1 + I_2 \tag{3.16}$$

$$I_A + I_C + I_1 + I_3 = 0 \tag{3.17}$$

$$I_1 = V_1.Y_1 = (V_A - V_B).Y_1 \tag{3.18}$$

$$I_2 = V_2.Y_2 = (V_A - V_B - V_C).Y_2 \tag{3.19}$$

$$I_3 = V_A.Y_3 \tag{3.20}$$

$$I = I_2 - I_C \tag{3.21}$$

$$Y_3 = 2Y_2 \tag{3.22}$$

where $Y_1 = jB_1 = j\omega C_1$, $Y_2 = jB_2 = j\omega C_2$ and $Y_3 = jB_3 = j\omega C_3$.

The symmetrical component equations given in Appendix A are written for a star-connected system as follows:

$$\begin{bmatrix} V_A \\ V_B \\ V_C \end{bmatrix} = \frac{1}{\sqrt{3}} \begin{bmatrix} 1 & 1 & 1 \\ 1 & h^2 & h \\ 1 & h & h^2 \end{bmatrix} \begin{bmatrix} V_0 \\ V_p \\ V_n \end{bmatrix} \tag{3.23}$$

$$\begin{bmatrix} I_A \\ I_B \\ I_C \end{bmatrix} = \frac{1}{\sqrt{3}} \begin{bmatrix} 1 & 1 & 1 \\ 1 & h^2 & h \\ 1 & h & h^2 \end{bmatrix} \begin{bmatrix} I_0 \\ I_p \\ I_n \end{bmatrix} \tag{3.24}$$

where h is the complex operator $e^{j2\pi/3}$.

From (3.16) and (3.17),

$$I_A + I_B + I_C - I_2 + I_3 = 0 . \tag{3.25}$$

Using (3.19), (3.23) and (3.24), (3.25) can be rewritten as

$$\frac{3I_0}{\sqrt{3}} - (V_A - V_B - V_C).Y_2 + V_A.Y_3 = 0 . \tag{3.26}$$

From (3.22), (3.23) and (3.26), the following equation is obtained:

$$\frac{3I_0}{\sqrt{3}} + \frac{3V_0}{\sqrt{3}} \cdot Y_2 = 0 \tag{3.27}$$

Since $I_0 = V_0 \cdot Y_0$, where Y_0 is the zero-sequence admittance of the motor, (3.27) may be written as

$$V_0 \cdot (Y_0 + Y_2) = 0 . \tag{3.28}$$

The sum of admitttances Y_0 and Y_2 is non-zero, hence the zero-sequence voltage V_0 must vanish according to (3.28).

Hence, if the condition prescribed by (3.22) is satisfied, zero-sequence voltage and current are absent in the SMIG. There are thus no zero-sequence losses, and the phase imbalance is contributed solely from negative-sequence quantities.

Equations (3.15) to (3.22) can now be solved in association with (3.23) and (3.24) to yield the positive-sequence and negative-sequence voltage components:

$$V_p = \frac{\sqrt{3}V}{h(1-h)} \cdot \frac{hY_n - 2Y_2 - (1-h)Y_1}{Y_p + Y_n + 3Y_1 + 2Y_2} \tag{3.29}$$

$$V_n = \frac{\sqrt{3}V}{h(1-h)} \cdot \frac{(1-h^2)Y_1 + 2Y_2 - h^2Y_p}{Y_p + Y_n + 3Y_1 + 2Y_2} \tag{3.30}$$

where Y_p and Y_n are the positive- and negative-sequence admittances of the three-phase IG, respectively.

For a given single-phase grid voltage V and speed (or per-unit slip), Y_p and Y_n are known and both V_p and V_n can be computed. The currents I_p and I_n can then be calculated from the positive- and negative-sequence equivalent circuits. The generator performance, such as phase voltages, phase currents, electromagnetic torque, power factor and efficiency, can subsequently be obtained.

3.3.3   Balanced Operation

Susceptances for perfect phase balance. It is of interest to investigate the values of susceptances that will result in balanced operation in the three-phase induction machine. Since negative-sequence voltage is absent when the generator is balanced, one obtains, from (3.30),

$$(1 - h^2)Y_1 + 2Y_2 - h^2Y_p = 0. \tag{3.31}$$

Assuming Y_1 and Y_2 to be pure *capacitive* admittances, i.e. $Y_1 = jB_1$ and $Y_2 = jB_2$,

(3.31) may be rewritten as two simultaneous algebraic equations:

$$\frac{1}{2}B_1 + B_2 = -\frac{Y_p}{\sqrt{3}}\cos\phi_p \tag{3.32}$$

$$\frac{3}{2}B_1 + B_2 = Y_p \sin\phi_p \tag{3.33}$$

where ϕ_p is the positive-sequence impedance angle of the IG.

The capacitive susceptances that result in perfect phase balance are obtained by solving (3.32) and (3.33):

$$B_1 = \frac{2}{\sqrt{3}}Y_p \sin\left(\frac{5\pi}{6} - \phi_p\right) \tag{3.34}$$

$$B_2 = Y_p \sin\left(\phi_p - \frac{2\pi}{3}\right) \tag{3.35}$$

$$B_3 = 2Y_p \sin\left(\phi_p - \frac{2\pi}{3}\right). \tag{3.36}$$

The values of phase converter susceptances required thus depend on Y_p and ϕ_p which are both functions of the rotor speed. Depending on the generator impedance angle, one or more of the susceptances may assume negative values, implying that inductances may have to be used for perfect phase balance. Table 3.4 summarizes the nature of the phase converter susceptances for different values of ϕ_p. When ϕ_p lies between $2\pi/3$ and $5\pi/6$ rad, B_1, B_2 and B_3 all have positive values, implying that perfect phase balance can be achieved by using capacitances only. When ϕ_p is less than $2\pi/3$ rad, B_1 is positive but B_2 and B_3 are negative, hence one capacitance and two inductances are required for perfect phase balance.

It is interesting to note that, when $\phi_p = 2\pi/3$ rad, B_2 and B_3 are both equal to zero, hence the capacitances C_2 and C_3 are not required. Under this condition, the Smith connection is identical to the Steinmetz connection for a star-connected IG. When $\phi_p = 5\pi/6$ rad, however, B_1 is equal to zero and only capacitances C_2 and C_3 need to be used for achieving perfect phase balance.

Table 3.4 *Susceptances for perfect phase balance in three-phase IG. Reproduced by permission of T.-F. Chan and L. L. Lai, 'Single-phase operation of a three-phase induction generator with the Smith connection,' IEEE Transactions on Energy Conversion, 17: 2002, 47–54. © (2002) IEEE*

Range of ϕ_p	B_1	B_2	B_3
$\phi_p < 2\pi/3$	Capacitive	Inductive	Inductive
$\phi_p = 2\pi/3$	Capacitive	Zero	Zero
$2\pi/3 < \phi_p < 5\pi/6$	Capacitive	Capacitive	Capacitive
$\phi_p = 5\pi/6$	Zero	Capacitive	Capacitive
$\phi_p > 5\pi/6$	Inductive	Capacitive	Capacitive

Line current and power factor. Referring to the phasor diagram of the SMIG shown in Figure 3.13(b), the following angular relationships may be deduced:

$$\alpha = 5\pi/6 - \phi_p \qquad \beta = \phi_p - 2\pi/3; \qquad \varepsilon = 5\pi/6;$$

$$\delta = \phi_p + \pi/6; \qquad \xi = \phi_p - 5\pi/6.$$

From the current phasor triangles, it can be shown that the line current I and the generator phase current I_{ph} are related by

$$I = I_{ph}\sqrt{1 + 8\sin^2\left(\phi_p - \frac{2\pi}{3}\right)}. \tag{3.37}$$

The angle γ between I and I_C in Figure 3.13(b) is given by

$$\gamma = \sin^{-1}\left[\frac{\sin 2(\phi_p - 2\pi/3)}{\sqrt{1 + 8\sin^2(\phi_p - 2\pi/3)}}\right]. \tag{3.38}$$

If the *input* power factor angle ϕ is defined to be positive when the line current I lags the supply voltage V, then

$$\phi = \delta + \gamma = \phi_p + \frac{\pi}{6} + \gamma. \tag{3.39}$$

Equations (3.38) and (3.39) indicate that the input power factor angle of the SMIG under perfect phase balance is a function of the generator impedance angle only. Figure 3.14 shows the variation of the line power factor angle ϕ with the positive-sequence impedance angle ϕ_p. It is observed that the line power factor angle is $180°$e (electrical degrees) (i.e. line power factor is equal to unity) when ϕ_p is equal to $130.89°$e. At higher values of ϕ_p, the SMIG delivers power at lagging power factor to the single-phase grid.

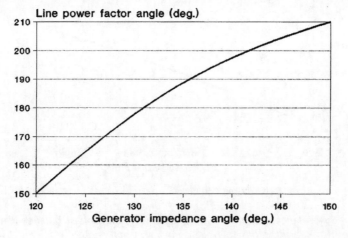

Figure 3.14   *Variation of line power factor angle with generator impedance angle*

3.3.4   Case Study

To verify the above analysis, a phase balancing experiment was performed on the machine IG1 (reconnected as the SMIG) whose parameters are given in Appendix D.1. With the single-phase grid voltage maintained constant at 380V, the rotor speed of the SMIG was controlled by a separately excited DC motor and the capacitances were varied until the phase voltages and currents were balanced. Figure 3.15 shows the computed and experimental values of susceptances B_1 and B_2 that give perfect balanced operation in the three-phase machine. For speeds above 1539 r/min, both B_1 and B_2 are positive (i.e. capacitive). As the rotor speed increases, B_1 decreases slightly while B_2 increases almost linearly, meaning that more and more reactive power is being furnished by the capacitances C_2 and C_3. Good correlation between the experimental and computed performance characteristics is observed. The results confirm that perfect phase balance in the three-phase IG can be achieved at different speeds using the Smith connection.

A load test was next conducted on the experimental machine with the following values of capacitances: $C_1 = 27\mu F$, $C_2 = 16\mu F$ and $C_3 = 32\mu F$. These capacitances enabled the IG to be balanced at a rotor speed of 1568 r/min and at a phase current of 5.25 A (0.97 of rated value). For convenience, this speed will be denoted by N_b in the subsequent discussion. Due to the effect of negative-sequence torque, the 'cut-in' speed N_c (i.e. speed at which the SMIG starts to deliver power to the grid) was 1515 r/min, which is considerably higher than the synchronous speed. Figures 3.16–3.20 show the experimental and computed performance characteristics of the SMIG. Figure 3.16 shows that, with fixed values of capacitances, the A-phase voltage increases as the speed decreases from N_b, while both the B-phase and C-phase voltages decrease. At speeds above N_b, overvoltage occurs

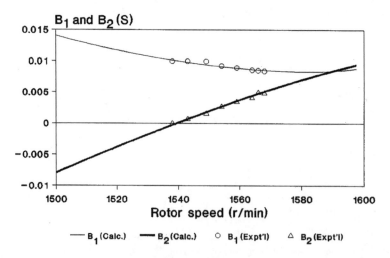

Figure 3.15   *Susceptances of phase converters to give perfect phase balance in experimental SMIG ($B_3 = 2\,B_2$). Reproduced by permission of T.-F. Chan and L. L. Lai, 'Single-phase operation of a three-phase induction generator with the Smith connection,' IEEE Transactions on Energy Conversion, **17**: 2002, 47–54. © (2002) IEEE*

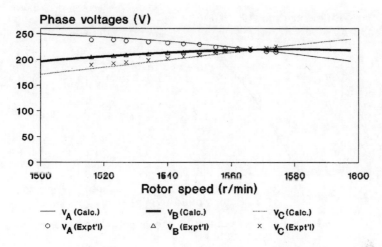

Figure 3.16 *16 Phase voltage variations of SMIG. Reproduced by permission of T.-F. Chan and L. L. Lai, 'Single-phase operation of a three-phase induction generator with the Smith connection,' IEEE Transactions on Energy Conversion, **17**: 2002, 47–54. © (2002) IEEE*

in C-phase. The voltage stress in A-phase and C-phase thus imposes a limit on the speed range over which the given set of capacitances could be left in the circuit. If the phase voltage is not to exceed 110 % of the rated value, then satisfactory generator operation from N_c up to 1600 r/min is possible.

Figure 3.17 shows the variation of phase and line currents of the SMIG with speed. Both I_A and I_C exhibit a concave-upward characteristic, but I_A is much higher at low speeds. At synchronous speed, I_A is equal to 7.1 A, or 131 % of the rated value. Starting from 1520 r/min, I_C increases very rapidly, eventually

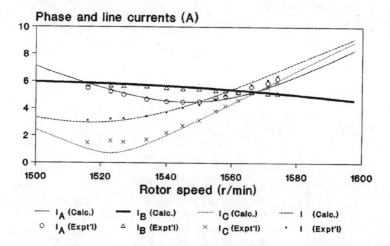

Figure 3.17 *Phase and line current variations of SMIG. Reproduced by permission of T.-F. Chan and L. L. Lai, 'Single-phase operation of a three-phase induction generator with the Smith connection,' IEEE Transactions on Energy Conversion, **17**: 2002, 47–54. © (2002) IEEE*

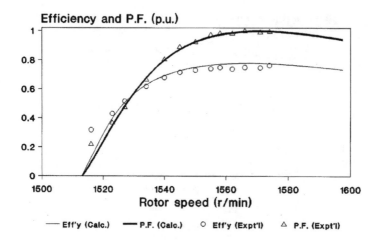

Figure 3.18   *Efficiency and power factor characteristics of SMIG. Reproduced by permission of T.-F. Chan and L. L. Lai, 'Single-phase operation of a three-phase induction generator with the Smith connection,' IEEE Transactions on Energy Conversion, **17**: 2002, 47–54. © (2002) IEEE*

exceeding both I_A and I_B for speeds above N_b. On the contrary, the B-phase current is relatively insensitive to changes in rotor speed. At speeds below 1540 r/min, I_B is almost constant at 6A (110 % of rated value).

Figure 3.18 shows the efficiency and power factor characteristics of the SMIG while Figure 3.19 shows the output power and driving torque characteristics. At N_b, the efficiency is very close to the maximum value of 0.78 p.u. and the SMIG delivers a power of 2320 W to the single-phase grid at approximately unity power factor. High efficiency and power factor are observed for speeds above 1540 r/min.

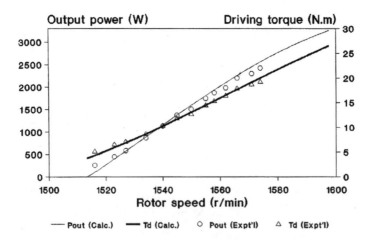

Figure 3.19   *Output power and driving torque characteristics of SMIG. Reproduced by permission of T.-F. Chan and L. L. Lai, 'Single-phase operation of a three-phase induction generator with the Smith connection,' IEEE Transactions on Energy Conversion, **17**: 2002, 47–54. © (2002) IEEE*

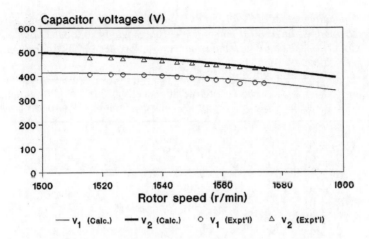

Figure 3.20 *Capacitor voltage variations of SMIG. Reproduced by permission of T.-F. Chan and L. L. Lai, 'Single-phase operation of a three-phase induction generator with the Smith connection,' IEEE Transactions on Energy Conversion, **17**: 2002, 47–54. © (2002) IEEE*

Figure 3.20 shows the variation of voltages across capacitances C_1 and C_2. Both V_1 and V_2 increase with reduction in rotor speed. At synchronous speed, V_1 reaches 135 % of rated value, while V_2 is equal to 109 % of rated value. The voltage ratings of the capacitances should thus be properly chosen in order to withstand the overvoltages at light load.

Good agreement between the experimental and theoretical results is observed in Figures 3.16–3.20, showing that the steady-state performance of the single-phase SMIG can be accurately determined using the method of symmetrical components.

Despite the overcurrent in B-phase and A-phase (at low speeds), the total losses in the SMIG are less than those corresponding to N_b, even at speeds down to N_c. The rotor copper loss may exceed the rated value at low speeds, due to the more predominant negative-sequence voltage. To avoid excessive temperature rise in the rotor, the SMIG should not be left in the 'idling' mode for a prolonged period of time.

3.3.5 Effect of Phase Balancing Capacitances

The effects of the phase balancing capacitances on generator performance will be discussed in the following sections with reference to the experimental machine. Referring to Table 3.5, Mode L, Mode M and Mode H denote operation with capacitances that result in perfect phase balance at light load, medium load and heavy load, respectively. For comparison, single-phasing operation (Mode 0) and operation with perfect phase balance (Mode V) are also investigated.

Output power. Figure 3.21 shows the computed output power characteristics of the SMIG. The speed N_c at which the generator begins to deliver power varies slightly depending on the phase balancing capacitances being used. At 1568 r/min, the power outputs of the SMIG for various modes are approximately equal, except

Table 3.5 *Operating modes of SMIG and the corresponding phase balancing capacitances. Reproduced by permission of T.-F. Chan and L. L. Lai, 'Single-phase operation of a three-phase induction generator with the Smith connection,' IEEE Transactions on Energy Conversion, **17**: 2002, 47–54. © (2002) IEEE*

Mode	Capacitances (μF)	Speed at perfect phase balance (r/min)	Power output at perfect balance (W)
0	$C_1 = 0$ $C_2 = 0$ $C_3 = 0$	n.a.	n.a.
L	$C_1 = 32.0$ $C_2 = 0$ $C_3 = 0$	1539	1235
M	$C_1 = 29.1$ $C_2 = 7.7$ $C_3 = 15.4$	1553	1750
H	$C_1 = 27.0$ $C_2 = 17.5$ $C_3 = 35.0$	1571	2430
V	Variable	Variable	Variable

for Mode 0 (single-phasing mode). The diagram also shows that, with the phase balancing capacitances, the IG can deliver 40 % more power than the single-phasing mode of operation at rated speed.

Power factor and efficiency. Figure 3.22 shows the power factor characteristics of the single-phase SMIG. With single-phasing mode, the power factor is very low, reaching only 0.5 (leading) at a speed of 1570 r/min. With Mode L capacitances, the power factor is much higher, exceeding 0.8 (leading) at speeds above 1525 r/min.

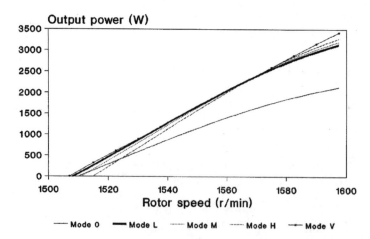

Figure 3.21 *Effect of capacitances on output power. Reproduced by permission of T.-F. Chan and L. L. Lai, 'Single-phase operation of a three-phase induction generator with the Smith connection,' IEEE Transactions on Energy Conversion, **17**: 2002, 47–54. © (2002) IEEE*

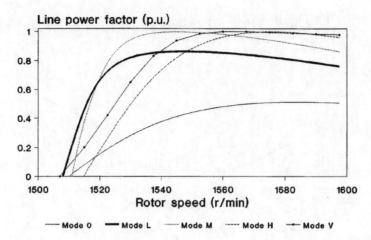

Figure 3.22   *Effect of capacitances on line power factor. Reproduced by permission of T.-F. Chan and L. L. Lai, 'Single-phase operation of a three-phase induction generator with the Smith connection,' IEEE Transactions on Energy Conversion, **17**: 2002, 47–54. © (2002) IEEE*

It is interesting to find that with Mode H capacitances, the power factor reaches unity at 1575 r/min. Below this speed, the power factor is lagging, implying that lagging reactive power is supplied to the grid. A similar trend is observed for Mode M, but unity power factor now occurs at a lower speed. For Mode V, the line power factor angle is constrained by the positive-sequence impedance angle of the IG (Figure 3.14), and power is delivered at a leading power factor when the rotor speed is below 1565 r/min.

The phase balancing capacitances are effective in suppressing the negative-sequence losses, with a consequent improvement in generator efficiency over the single-phasing mode of operation. As illustrated in Figure 3.23, the efficiencies

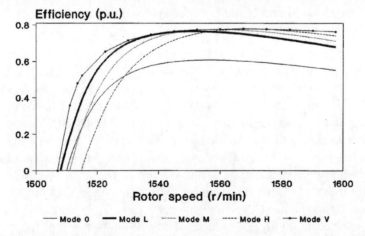

Figure 3.23   *Effect of capacitances on efficiency. Reproduced by permission of T.-F. Chan and L. L. Lai, 'Single-phase operation of a three-phase induction generator with the Smith connection,' IEEE Transactions on Energy Conversion, **17**: 2002, 47–54. © (2002) IEEE*

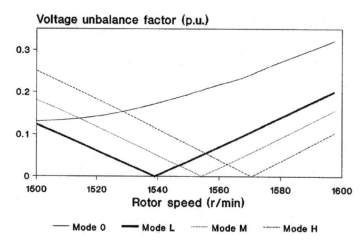

Figure 3.24 *Effect of capacitances on VUF. Reproduced by permission of T.-F. Chan and L. L. Lai, 'Single-phase operation of a three-phase induction generator with the Smith connection,' IEEE Transactions on Energy Conversion, **17**: 2002, 47–54. © (2002) IEEE*

for Modes L, M and H are maximum when the generator is balanced, i.e. when the corresponding characteristic touches the Mode V curve. With the generator balanced at a large power output, however, the efficiency at low speeds is reduced due to a more severe phase imbalance.

Voltage unbalance factor. The degree of phase imbalance can conveniently be described in terms of the voltage unbalance factor (VUF), which is the ratio of the negative-sequence voltage V_n to the positive-sequence voltage V_p [4, 2]. As shown in Figure 3.24, the VUF for single-phasing mode increases monotonically with speed, whereas the VUF characteristics for Modes L, M and H are V-shaped curves, each with a minimum value of zero at the respective balance points. It is observed that Mode M and Mode H capacitances produce a larger VUF than the single-phasing mode at low speeds.

Electromagnetic torque. Figure 3.25 shows the variations of electromagnetic torque of the single-phase SMIG with per-unit slip (absolute value). With Modes L, M and H, the torque–slip characteristics are very close to that for Mode V over the normal slip range (i.e. from 0 to 0.06). At higher values of per-unit slip, the positive-sequence voltage component decreases significantly, resulting in a smaller pull-out torque and a narrower stable operating region. It is observed that the pull-out torques for Modes L, M and H occur at values of per-unit slip from 0.12 to 0.14, compared with 0.25 for Mode V. The pull-out torque for Mode H is 38 N m, which is 1.9 times the rated torque. The pull-out torque for Mode V is much higher, but large values of capacitance C_1 and inductances L_2 and L_3 are required for achieving perfect phase balance at the pull-out slip. The pull-out torque for Mode L is larger than that for Mode M and Mode H, despite the fact that the values of capacitances used are smaller.

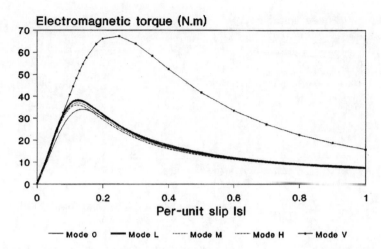

Figure 3.25 *Effect of capacitances on electromagnetic torque. Reproduced by permission of T.-F. Chan and L. L. Lai, 'Single-phase operation of a three-phase induction generator with the Smith connection,' IEEE Transactions on Energy Conversion, **17**: 2002, 47–54. © (2002) IEEE*

3.3.6 Dual-Mode Operation

A compromise has to be made between the permissible voltage imbalance and the cost in providing the additional phase balancing capacitances and the associated switches. From economic considerations, a simple dual-mode control scheme involving only two sets of capacitances, e.g. Mode L and Mode H, may be employed. Referring to Figure 3.24, the VUF characteristics of the experimental machine with Mode L and Mode H capacitances intersect at a speed N_{sw} equal to 1556 r/min. This speed demarcates the Mode L and Mode H operating regions for the proposed dual-mode control scheme. For speeds below N_{sw}, Mode L capacitances are used since they yield a smaller VUF. For speeds above N_{sw}, Mode H capacitances should be used. Figure 3.26 shows the voltage variations of the experimental single-phase SMIG with dual-mode control. From N_{sw} to 1590 r/min and with Mode H capacitances in the circuit, the maximum overvoltage occurs in C-phase and is equal to 13 V (5.9 % of rated value). At speeds below N_{sw}, the maximum overvoltage occurs in A-phase and is equal to 15 V (6.8 % of rated value). Also shown in Figure 3.26 is the variation in the VUF, which is now a W-shaped curve. Perfect phase balance is obtained at 1539 r/min at which the SMIG is delivering half load and again at 1571 r/min at which the SMIG is delivering full-load power output. At the switching speed N_{sw}, the VUF is equal to 0.056 and the maximum overvoltage is less than 11 V (5 % of rated value).

Figure 3.27 shows the phase and line current characteristics of the SMIG with dual-mode control. From synchronous speed up to 1553 r/min and with Mode L capacitances in the circuit, all the phase currents are less than the rated value. Above this speed, a slight overcurrent occurs in C-phase, reaching 6 % of the rated value at the switching speed N_{sw}. After changeover to Mode H capacitances, a small

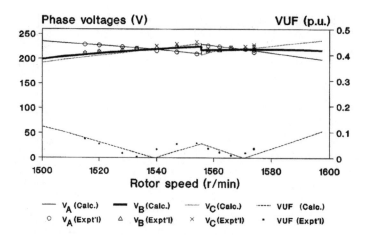

Figure 3.26   *Phase voltages and VUF of SMIG with dual-mode control. Reproduced by permission of T.-F. Chan and L. L. Lai, 'Single-phase operation of a three-phase induction generator with the Smith connection,' IEEE Transactions on Energy Conversion, **17**: 2002, 47–54.* © *(2002) IEEE*

overcurrent occurs in B-phase, but both A-phase and C-phase currents are now less than the rated value.

The experimental results shown in Figure 3.26 and Figure 3.27 suggest that satisfactory operation of the single-phase SMIG over a wide range of power output can be achieved by using a simple dual-mode control scheme.

3.3.7   Summary

The feasibility of a grid-connected single-phase IG system based on the Smith connection has been demonstrated in this section. A systematic analysis using

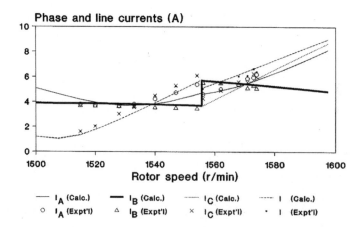

Figure 3.27   *Phase and line currents of SMIG with dual-mode control. Reproduced by permission of T.-F. Chan and L. L. Lai, 'Single-phase operation of a three-phase induction generator with the Smith connection,' IEEE Transactions on Energy Conversion, **17**: 2002, 47–54.* © *(2002) IEEE*

the method of symmetrical components has been presented for evaluation of the generator performance at different rotor speeds. The interesting case of balanced operation in the three-phase induction machine has been investigated in detail, and the conditions necessary for perfect phase balance are deduced. It is shown that, when the generator impedance angle is between $2\pi/3$ and $5\pi/6$ rad, phase balancing can be achieved by using capacitances only. The effect of capacitances on the generator performance has also been investigated. From a consideration of the VUF, a simple dual-mode control strategy that gives satisfactory generator performance over a wide range of power output is proposed. The theoretical analysis is validated by experiments performed on a small induction machine.

3.4 Microcontroller-Based Multi-Mode Control of SMIG

3.4.1 Phase Voltage Consideration

The theoretical analysis and experimental results presented in Section 3.3 reveal the following facts about the SMIG:

1. For a given set of capacitances, the phase voltages and phase currents will change when the rotor speed changes. The A-phase voltage V_A decreases with increase in rotor speed, whereas the B-phase voltage V_B and C-phase voltage V_C both increase.
2. The greater the deviation of rotor speed from that corresponding to perfect balance, the greater the deviation of the phase voltages from the value corresponding to perfect phase balance.
3. V_C is more sensitive to rotor speed change than V_B.
4. V_C varies almost linearly with the rotor speed.

The above observations suggest that, instead of the rotor speed signal, the C-phase voltage V_C may be exploited for controlling the switching of capacitances for multi-mode operation of the IG. An advantage of this approach is that an expensive speed sensor need not be used.

3.4.2 Control System

For practical design, a compromise has to be made between the phase imbalance permissible and the cost of the controller system. Previous studies have indicated that a dual-mode or three-mode controller will suffice to yield satisfactory machine performance. In this section, the implementation of a three-mode controller will be described. For easy reference, the three modes are referred to as L-mode (low load), M-mode (medium load) and H-mode (heavy load), respectively.

To reduce circuit complexity and to improve system reliability, a digital approach based on microcontroller technology has been adopted. A microcontroller may be viewed as a compact computer manufactured on a single chip. The built-in

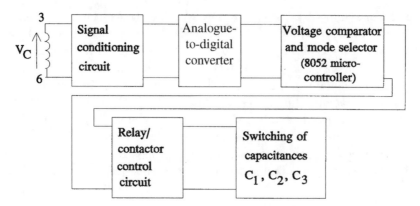

Figure 3.28   *Schematic diagram of proposed microcontroller-based mode selection system*

input/output (I/O) and memory systems enable the chip to be interfaced with the hardware system to be controlled. At present, microcontrollers of the 8051 family [8] are widely used for implementing stand-alone, embedded control systems.

Figure 3.28 shows a schematic diagram of the proposed controller. The C-phase voltage V_C is input to a signal conditioning circuit which consists of a two-winding step-down transformer, a diode rectifier and a sample-and-hold circuit. The sampled DC signal is next input to an analogue-to-digital converter (ADC). The digital control signal is input to an 8052 microcontroller (an enhanced member of the 8051 family) which functions as a voltage comparator and a mode selector. The control functions are executed by an assembly language program that has previously been developed, compiled and linked using an assembler, and 'burned in' the read-only memory (ROM) of the microcontroller chip [9]. The output from the microcontroller is then used to drive a relay/contactor circuit that effects capacitor switching. The system cost is low because of the well-established microcontroller technology and the small component count.

It should be noted that the 8052 chip cannot be programmed by the user. In other words, the control program has to be supplied to the manufacturer for producing the microcontroller. This approach is not economical for prototyping, so in practice an enhanced version of the chip with erasable and programmable read-only memory (EPROM) will be used. Program development and debugging can now be undertaken under the personal computer (PC) environment and the final program can be written into the EPROM.

The proposed control strategy can be explained by referring to Figure 3.29 which shows the variation of V_C with rotor speed for the three sets of capacitances. For this example, the balance points are LB, MB and HB, at which V_C is equal to the rated value of 130 V. The threshold value ΔV has been assigned to be ± 5 V as indicated by the two dashed horizontal lines at 125 V and 135 V. The IG is allowed to operate in its present mode provided that the change in V_C does not exceed the specified voltage threshold. A mode change or an alarm will be initiated, however,

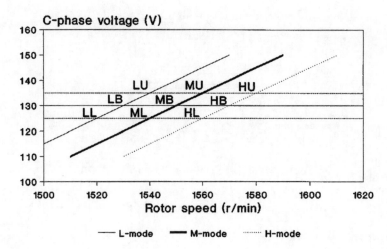

Figure 3.29   *Variation of C-phase voltage with speed for L-mode, M-mode and H-mode capacitances*

when the change in V_C reaches or exceeds the threshold value. The latter provision is necessary as the phase imbalance will be too large.

Assume that the IG is initially operating at perfect phase balance with L-mode capacitances in the circuit. As the rotor speed increases, V_C will vary according to the L-mode characteristic (thin solid curve in Figure 3.29) from point LB towards LU. At point LU the threshold voltage is reached and a mode change is initiated. M-mode capacitances are switched into the circuit and V_C will then vary according to the M-mode characteristic (thick solid curve in Figure 3.29). Further increase in rotor speed will cause V_C to reach point MU at which the H-mode capacitances are switched in. If the rotor speed continues to increase, the point HU on the H-mode characteristic (dotted curve in Figure 3.29) will finally be reached and further mode change is not possible. An alarm signal will be issued to caution the operator for appropriate action.

Change of mode in the reverse sequence takes place in a similar manner when the rotor speed decreases. The transition from H-mode to M-mode occurs at point HL, while the transition from M-mode to L-mode occurs at point ML. For proper mode change operation, the voltage threshold should be chosen such that, in Figure 3.29, HL is on the left of MU along the speed axis and ML is on the left of LU.

Figure 3.30 shows a flowchart of the voltage comparison routine of the Assembly program in the microcontroller. The function of this routine is to keep track of the present operating mode of the IG and to determine whether a mode change is necessary. The program reads the V_C signal from the ADC continuously and compares it against the following voltages:

V_b = voltage of V_C at perfect balance;
V_L = minimum value of V_C permissible;
V_U = maximum value of V_C permissible.

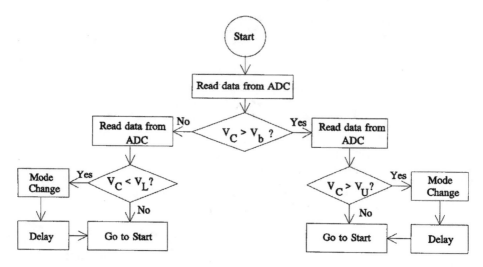

Figure 3.30   *Flowchart of the voltage comparison routine of the control program*

As long as the condition $V_L < V_C < V_U$ prevails, no action is taken and the controller continues to read the sampled V_C signal. When $V_C > V_U$ or $V_C < V_L$, however, a mode change will be initiated. To avoid rapid mode changes back and forth due to transient disturbances, the sampled signal is read twice before initiating a mode change and a time delay is introduced after each mode change.

Figure 3.31 shows a flowchart of the mode selection routine. When the 'Mode change' command is issued from the voltage comparison routine, the present operating mode is recalled and the appropriate mode is selected accordingly. An alarm signal will be issued when no further mode change is possible.

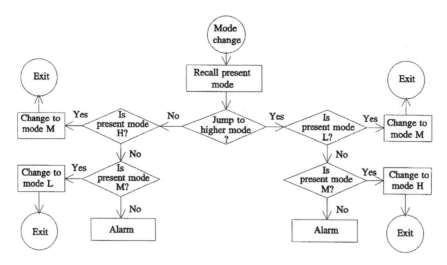

Figure 3.31   *Flowchart of mode selection routine of the control program*

Table 3.6 *Capacitances for perfect phase balance in experimental machine*

Mode	C_1 (μF)	C_2 (μF)	C_3 (μF)	Rotor speed at balance (r/min)
L	20	10	20	1533
M	15	25	50	1552
H	12.5	40	80	1574

3.4.3 Practical Implementation

Tests performed on a 50 Hz, four-pole, 130 V (phase), 5.2 A (phase) experimental machine gave the sets of capacitances for perfect phase balance as shown in Table 3.6. Based on these results, the capacitor switching arrangement shown in Figure 3.32 was designed and the Assembly language control program was developed. Table 3.7 shows how the six switches S1 to S6 should be controlled in order to give the capacitance values in Table 3.6. The system would be initialized to L-mode immediately after the controller was energized. Thereafter the appropriate mode would be selected based on the measured V_C signal.

The 89C52 microcontroller from ATMEL, which is one of the enhanced members of the 8052 family, was selected for hardware implementation. This chip is a high-performance CMOS 8-bit microcontroller with 8 kbytes of EPROM and 256 kbytes of random access memory (RAM), and is compatible with the 8051/8052 instruction set. There are four parallel I/O ports. In the prototype system, port P1 was used for digital signal input from the ADC. Port P0 was used for control of the relay/contactor circuit, while port P2 was used for control of the mode and alarm indication circuit.

Calibration of the sampled DC signal against the C-phase voltage V_C was necessary for software program development. Figure 3.33 shows the calibration curve obtained on the experimental system. Due to the voltage drop in the diode rectifier of the signal conditioning circuit, nonlinearities occur when V_C is small. Over the

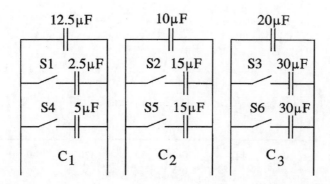

Figure 3.32 *Capacitor switching arrangement of single-phase IG with the Smith connection*

Table 3.7   *Switch control for phase balancing capacitances*

Mode	C_1	C_2	C_3
L	S1 closed	S2 open	S3 open
	S4 closed	S5 open	S6 open
M	S1 closed	S2 closed	S3 closed
	S4 open	S5 pen	S6 open
H	S1 open	S2 closed	S3 closed
	S4 open	S5 closed	S6 closed

practical operating range (from 120 to 140 V), however, the sampled DC signal is approximately proportional to V_C.

Heavy-current relays were used for switches S1 to S6 in the prototype system. Voltage amplification of the digital output from the microcontroller was required for energizing the relays. For larger machines, the use of zero-crossing solid-state relays [10] is preferred in order to minimize the transient switching currents.

3.4.4   Experimental Results

Figure 3.34 shows the phase voltage variations obtained from a load test on the experimental SMIG system with the multi-mode controller, the rotor speed being increased monotonously from synchronous value (1500 r/min). It is observed that the phase voltages vary within close limits about the rated value (130 V) throughout the normal speed range. Switching over from L-mode to M-mode occurs at a rotor

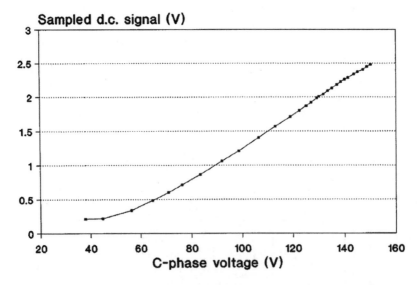

Figure 3.33   *Variation of the sampled DC signal against V_C*

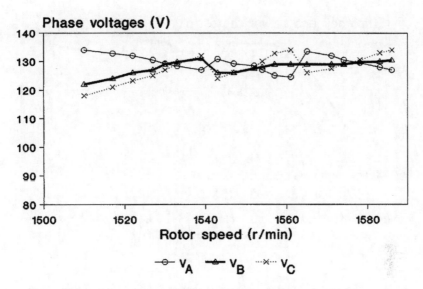

Figure 3.34 *Phase voltage variations of SMIG under multi-mode operation*

speed of 1541 r/min, and switching from M-mode to H-mode occurs at a rotor speed of 1563 r/min. Figure 3.34 has further confirmed that V_C is an appropriate control signal for mode selection.

Figure 3.35 shows the corresponding variations of the phase and line currents of the IG. For rotor speeds close to the balance points, the phase current imbalance is quite small. At very light loads, however, the imbalance becomes more severe. The

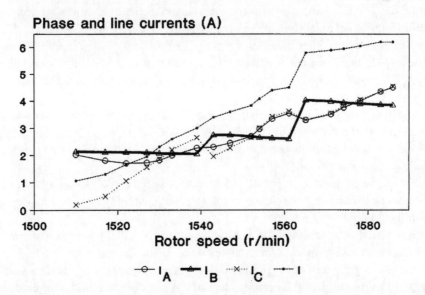

Figure 3.35 *Phase and line current variations of SMIG under multi-mode operation*

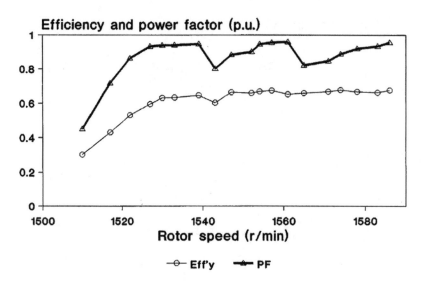

Figure 3.36   *Efficiency and power factor variations of SMIG under multi-mode operation*

phase currents under this condition, however, are much lower than the rated value (due to the use of L-mode capacitances) and hence the motor losses are acceptable. Within each mode, the B-phase current I_B decreases only slightly with increase in rotor speed, but its magnitude is larger when the next higher mode is selected. I_A and I_C, on the other hand, vary considerably within the same mode, but the two currents are approximately equal for operation in the H-mode.

Figure 3.36 shows the efficiency and power factor characteristics of the SMIG. Despite the mode changes, the efficiency characteristic is quite smooth and the efficiency is acceptable at rotor speeds above 1530 r/min. The power factor characteristic, on the other hand, exhibits marked discontinuities due to the mode changes. Over the practical speed range, however, the output line power factor exceeds 0.8 (leading) while at speeds corresponding to perfect phase balance, the line power factor is above 0.9 (leading).

Similar performance characteristics were obtained when the rotor speed was reduced from that corresponding to heavy load. Switching from H-mode to M-mode was found to occur at a rotor speed of 1567 r/min and switching from M-mode to L-mode at a rotor speed of 1542 r/min.

The response of the microcontroller to transient speed changes was also studied. For a moderate increase or decrease in rotor speed, switching between modes was accomplished in about 1.5 s.

Experimental waveforms of the SMIG were also recorded when the rotor speed was 1533 r/min and L-mode capacitances were in the circuit. Figures 3.37–3.39 show the oscillograms of the experimental waveforms obtained by using a voltage scale of 200 V per division, a current scale of 2 A per division, and a time scale of 5 ms per division. The waveform of the supply voltage V (the upper trace) in each

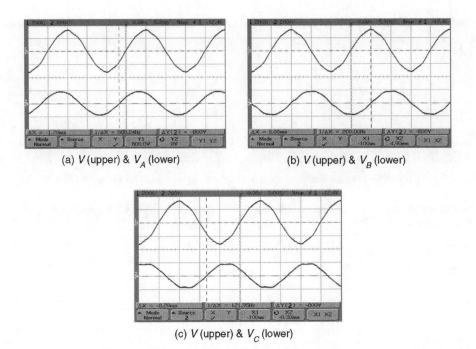

(a) V (upper) & V_A (lower)

(b) V (upper) & V_B (lower)

(c) V (upper) & V_C (lower)

Figure 3.37 *Phase voltage waveforms of SMIG: balanced operation with L-mode capacitances*

oscillogram provides a convenient reference for studying the phase relationship between various voltages and currents.

The oscillograms in Figures 3.37–3.39 show that the IG was operating under approximately balanced conditions. An examination of the phase angles (using the vertical cursors in the oscillograms) confirms the validity of the phasor diagram shown in Figure 3.13. The voltages and currents are approximately sinusoidal, but both the C-phase voltage V_C and the line current I are flat topped. These waveform distortions are caused by the local loops in the Smith connection which provide closed paths for the flow of current harmonics. For example, the harmonic components in V_C will circulate the corresponding harmonic currents through capacitances C_1 and C_2 in loop 3453 of Figure 3.12. Similarly, harmonic currents will flow in capacitance C_3 due to harmonics in the A-phase voltage V_A. This phenomenon is confirmed by the waveforms of capacitor currents shown in Figure 3.39. Third-harmonic distortion, for example, can be observed from the waveform of I_1 in Figure 3.39(a).

3.4.5 Summary

Microcontroller-based multi-mode operation of a grid-connected single-phase SMIG has been presented in this section. The control system and mode switching

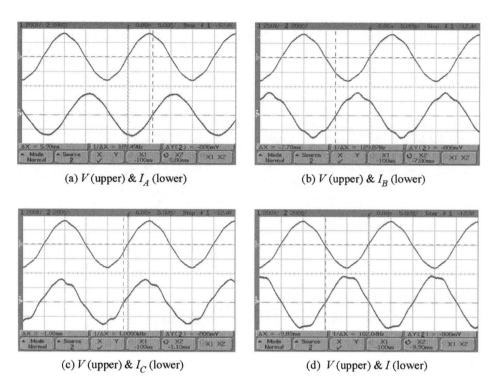

(a) V (upper) & I_A (lower)

(b) V (upper) & I_B (lower)

(c) V (upper) & I_C (lower)

(d) V (upper) & I (lower)

Figure 3.38   *Phase and line current waveforms of SMIG: balanced operation with L-mode capacitances*

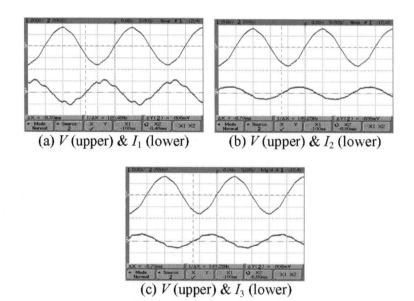

(a) V (upper) & I_1 (lower)

(b) V (upper) & I_2 (lower)

(c) V (upper) & I_3 (lower)

Figure 3.39   *Capacitor current waveforms of SMIG: balanced operation with L-mode capacitances*

strategy have been described. The prototype controller system implemented has confirmed the feasibility of the proposed design. Satisfactory generator performance has been obtained on a small laboratory induction machine. The experimental waveforms have verified the relationships between voltages and currents in the SMIG. This section has also demonstrated that a low-cost, reliable, practical single-phase IG system could be realized using the microcontroller approach.

3.5 Phase Balancing using a Line Current Injection Method

3.5.1 Circuit Connection and Operating Principle

Among the phase balancing schemes for IGs introduced by Smith [11], a novel line current injection method (namely, the Mode C circuit) deserves special attention. Whereas all the other circuits require a three-wire, two-level single-phase voltage supply for providing the injected current, the Mode C circuit requires only a two-wire single-phase supply (which is more often used universally) and employs a transformer for furnishing one of the injected line current components. Perfect phase balance could be achieved when the generator power factor angle is between $2\pi/3$ and $5\pi/6$ rad. In [11], the current injection transformer is assumed to have a unity primary/secondary turns ratio.

In this section, a systematic analysis on the above Mode C circuit for single-phase operation of a three-phase IG will be presented. The effect of non-unity transformer turns ratio on machine performance and the conditions of phase balance will also be investigated.

Figure 3.40 shows the Mode C circuit for single-phase operation of a three-phase delta-connected IG. A-phase (the reference phase) is connected across the

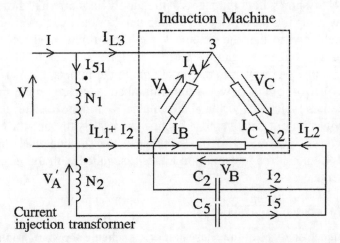

Figure 3.40 *Mode C circuit for single-phase operation of a three-phase delta-connected IG*

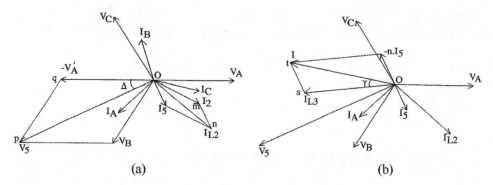

(a)   (b)

Figure 3.41   *Phasor diagrams for the Mode C circuit: (a) phasor diagram showing components of I_{L2}; (b) phasor diagram showing components of grid current I, the transformer no-load current having been neglected*

single-phase grid of voltage V while a capacitance C_2 is connected across B-phase (the lagging phase). The primary winding of the current injection transformer is connected across A-phase. The secondary winding voltage V_A', together with the B-phase voltage V_B, causes a current I_5 to flow through the capacitance C_5. The capacitor currents I_2 and I_5 thus constitute the line current I_{L2} that flows into terminal 2 of the IG.

Consider the phasor diagram in Figure 3.41(a), drawn for the special case of perfect phase balance. The line current I_{L2} lags the C-phase voltage V_C by $(\phi_p + \pi/6)$ rad, while the current I_2 lags V_C by $5\pi/6$ rad. Provided ϕ_p is greater than $2\pi/3$ rad, I_{L2} can always be formed from appropriate values of I_2 and I_5 and perfect phase balance can be obtained. The maximum value of generator impedance angle $\phi_{p,max}$ below which perfect phase balance is possible depends on the angle Δ which varies as the turns ratio of the current injection transformer.

Figure 3.41(b) shows the components of the supply line current I, the transformer no-load current having been neglected. Note that both active power and reactive power are delivered to the line voltage V via the current injection transformer (by virtue of the voltage V_A and the current component $-n.I_5$). The IG delivers balanced three-phase power from the stator winding terminals 1, 2 and 3 as shown in Figure 3.40. Most of this power is transmitted conductively to the single-phase line via terminals 1 and 3 (by virtue of the voltage V_A and the line current I_{L3}). The remaining power output is delivered to the line inductively through the transformer coupling mechanism. Figure 3.41(b) also suggests that the line power factor of the IG system is high and the transmission losses associated with the reactive power supply are reduced.

3.5.2   Performance Analysis

The performance analysis of the phase balancing scheme shown in Figure 3.40 can also be carried out using the method of symmetrical components [4]. Adopting the *motor* convention for the induction machine and neglecting the leakage impedance drop of the current injection transformer, the following inspection equations may be established:

$$V = V_A \tag{3.40}$$

$$V_A + V_B + V_C = 0 \tag{3.41}$$

$$I_2 = V_B . Y_2 \tag{3.42}$$

$$I_5 - (V_B - V'_A).Y_5 \tag{3.43}$$

$$I_{L3} = I_A - I_C \tag{3.44}$$

$$I_{L2} = I_2 + I_5 \tag{3.45}$$

$$I_{L2} = I_C - I_B \tag{3.46}$$

$$I = I_{L3} + I_{51} \tag{3.47}$$

where $Y_2 = j\omega C_2$, $Y_5 = j\omega C_5$ and ω is the angular frequency of the supply voltage.

The primary and secondary quantities of the current injection transformer are related by

$$I_{51} = I_{t0} - n.I_5 \tag{3.48}$$

$$V'_A = n.V_A \tag{3.49}$$

where n is the secondary-to-primary turns ratio N_2/N_1 and I_{t0} is the no-load current of the transformer.

Solution of the above equations using the method of symmetrical components gives the positive-sequence voltage V_p and the negative-sequence voltage V_n:

$$V_p = \sqrt{3}V \cdot \frac{Y_n + (Y_2 + Y_5)/(1 - h) - n.Y_5/(h - h^2)}{Y_p + Y_n + Y_2 + Y_5} \tag{3.50}$$

$$V_n = \sqrt{3}V \cdot \frac{Y_p - h.(Y_2 - Y_5)/(1 - h) + n.Y_5/(h - h^2)}{Y_p + Y_n + Y_2 + Y_5} \tag{3.51}$$

where Y_p and Y_n are the positive-sequence and negative-sequence admittances of the three-phase IG and h is the unit complex operator $e^{j2\pi/3}$.

For a given single-phase grid voltage V and speed (or per-unit slip), Y_p and Y_n are known and both V_p and V_n can be computed. The currents I_p and I_n can then be calculated from the positive-sequence and negative-sequence equivalent circuits.

The generator performance, such as phase voltages, phase currents, electromagnetic torque, power factor and efficiency, can subsequently be obtained.

3.5.3   Balanced Operation

Conditions for Balanced Operation.   Since negative-sequence voltage must be absent when the generator is balanced, one obtains, from (3.51),

$$Y_p - h.(Y_2 - Y_5)/(1 - h) + n.Y_5/(h - h^2) = 0. \tag{3.52}$$

Equation (3.52) may be solved to give the capacitive admittances that result in perfect phase balance. Alternatively, the phasor diagram under balanced conditions can be used. Referring to Figure 3.41(a), the voltage V_5 across C_5 is given by

$$V_5 = k_n.V_{ph} \tag{3.53}$$

where

$$k_n = \sqrt{1 + n + n^2} \tag{3.54}$$

and V_{ph} is the phase voltage.

The angle Δ in phasor triangle Oqp is given by

$$\sin \Delta = \frac{\sqrt{3}}{2.k_n}. \tag{3.55}$$

Applying the sine rule to the phasor triangle Omn yields

$$I_2 = \frac{\sqrt{3}I_{ph}.\sin(\pi - \Delta - \phi_p)}{\sin(2\pi/3 + \Delta)} \tag{3.56}$$

$$I_5 = \frac{\sqrt{3}I_{ph}.\sin(\phi_p - 2\pi/3)}{\sin(2\pi/3 + \Delta)} \tag{3.57}$$

where I_{ph} is the phase current and ϕ_p is the positive-sequence impedance angle of the IG. The admittances Y_2 and Y_5 that result in perfect phase balance are thus given by

$$Y_2 = \sqrt{3}Y_p.\frac{\sin(\pi - \Delta - \phi_p)}{\sin(2\pi/3 + \Delta)} \tag{3.58}$$

$$Y_5 = \frac{\sqrt{3}Y_p}{k_n}.\frac{\sin(\phi_p - 2\pi/3)}{\sin(2\pi/3 + \Delta)}. \tag{3.59}$$

Equation (3.58) implies that admittance Y_2 is positive when ϕ_p is less than $\pi - \Delta$. It is obvious that both Y_2 and Y_5 (and hence C_2 and C_5) depend on Y_p and ϕ_p, as well as the transformer turns ratio n.

Figure 3.42 shows the variations of Y_2 and Y_5 for balanced operation of the experimental IG whose technical data are given in Section 3.5.4. For a given value of ϕ_p, an increase in n results in an increase of Y_2 but a decrease of Y_5. This feature

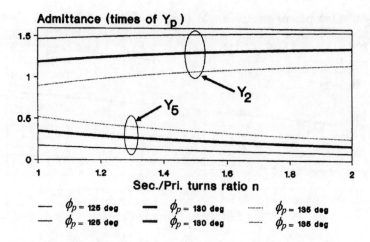

Figure 3.42   *Effect of transformer turns ratio n on the values of Y_2 and Y_5 for balanced operation of IG*

may be exploited in the design of a practical system. Use of values of n exceeding 2, however, is not recommended as the effect on Y_2 and Y_5 is less pronounced.

Line current and power factor. If the no-load current of the current injection transformer is neglected, the line current I and the angle γ can be computed from the phasor triangle Ost in Figure 3.41(b), as follows:

$$I = \sqrt{(n.I_5)^2 + I_{L3}^2 - 2\,(n.I_5)I_{L3}\cos(\phi_p - \pi/3 + \Delta)} \tag{3.60}$$

$$\sin\gamma = \frac{n.I_5}{I}.\sin(\phi_p - \pi/3 + \Delta). \tag{3.61}$$

If the *input* power factor angle ϕ is defined to be positive when the line current I lags the grid terminal voltage V, then

$$\phi = \phi_p + \frac{\pi}{6} + \gamma. \tag{3.62}$$

Figure 3.43 shows the variation of the line power factor angle ϕ as n is varied from unity to 2. For a given value of ϕ_p, ϕ increases almost linearly with n. When $\phi_p = 125°$e, ϕ is less than $180°$e and the *output* line power factor of the system is leading. The line power factor therefore becomes higher as n increases. When $\phi_p = 130°$e, the line power factor is very close to unity, but a transition from leading to lagging power factor occurs when $n = 1.2$. At larger values of ϕ_p, the system exports reactive power to the grid.

From (3.59), it is observed that when the turns ratio n is equal to unity and $\phi_p = 2\pi/3$ rad, Y_5 is equal to zero, hence the capacitance C_5 is not required and the circuit is reduced to the Steinmetz connection for a delta-connected IG. When

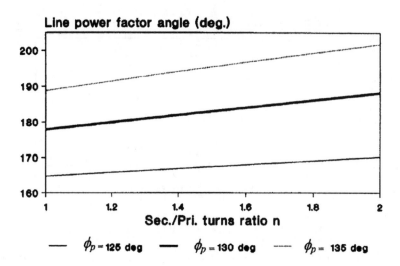

Figure 3.43   *Effect of transformer turns ratio n on the line power factor angle ϕ*

$\phi_p = 5\pi/6$ rad, however, Y_2 is equal to zero in accordance with (3.58). This is the maximum value of the generator positive-sequence impedance angle for which perfect phase balance can be achieved.

By choosing a value of n greater than unity and using the correct values of capacitances, balanced operation can be achieved for values of ϕ_p exceeding $5\pi/6$ rad. When n is increased to 1.2, for example, Y_2 is equal to zero when $\phi_p = 0.85\pi$ rad. This corresponds to a generator power factor of 0.89, achievable probably for a very large IG. In most cases, a current injection transformer with n close to unity would suffice for achieving balanced operation.

3.5.4   Case Study

Computation of Performance. To verify the above analysis, performance computations and experiments were performed on a 2.0 kW, 200V, 9.4 A, 50 Hz, four-pole, delta-connected induction machine. Figure 3.44 shows the variations of Y_p and ϕ_p with rotor speed, determined experimentally by operating the induction generator on a balanced three-phase supply at a constant voltage V_p^*. Figure 3.45 shows the corresponding variations of positive-sequence electromagnetic torque T_p^*. For a specific rotor speed, the positive-sequence torque was computed from the following equation:

$$T_p = T_p^* \times \left(V_p/V_p^*\right)^2 \tag{3.63}$$

where V_p is the computed positive-sequence voltage, while V_p^* and T_p^* are the corresponding values as given by Figure 3.44 and Figure 3.45.

The negative-sequence quantities were determined in a similar manner. For the experimental machine, the negative-sequence admittance Y_n was determined as $0.1\angle - 59°$ S and could be regarded as constant over the normal speed range. The

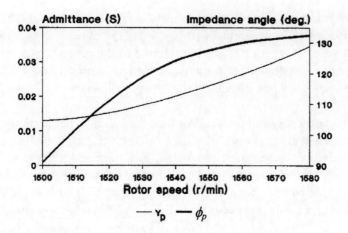

Figure 3.44 *Experimental variations of positive-sequence admittance Y_p and positive-sequence impedance angle ϕ_p with rotor speed, determined at $V_p^* = 200\,V$*

negative-sequence electromagnetic torque T_n was also insensitive to the rotor speed but depended on the applied voltage. For a given rotor speed, T_n was computed as

$$T_n = T_n^* \times \left(V_n/V_n^*\right)^2 \tag{3.64}$$

where T_n^* is the negative-sequence torque determined with an applied negative-sequence voltage of V_n^*. For the experimental machine, $T_n^* = 1.9\,\mathrm{N\,m}$ when $V_n^* = 54\,V$.

The driving torque T_d of the IG was computed as

$$T_d = T_p + T_n + T_{fw} \tag{3.65}$$

where T_{fw} is the torque for overcoming the friction and windage losses.

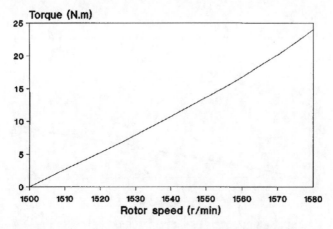

Figure 3.45 *Experimental variation of positive-sequence electromagnetic torque of IG with rotor speed, determined at $V_p^* = 200\,V$*

The current injection transformer used was rated at 50 Hz, 200/210.5 V and 1 kVA. It had a no-load current of $(0.107 - j0.259)$ A and an equivalent leakage impedance of $(1.4 + j2.89)$ Ω, both referred to the primary side. The leakage impedance drop was neglected in the performance computations, but the copper losses were included when evaluating the system efficiency.

Computed and experimental results. A phase balancing experiment was next performed at different rotor speeds. The grid voltage was maintained at rated value and the capacitances were varied until balanced generator operation was obtained. Figure 3.46 shows the computed and experimental values of capacitances C_2 and C_5 that result in balanced operation of the three-phase IG. It is observed that perfect phase balance is possible for speeds above 1533 r/min. As the rotor speed increases, C_5 increases approximately linearly while C_2 remains substantially constant over most of the rotor speed range.

Figure 3.47 shows the variations of phase current, line current and output power when the IG operates under perfect phase balance. At a speed of 1562.5 r/min (which corresponds to operation with approximately rated phase current), the generator delivers a power of 2.09 kW to the single-phase grid. Due to the injected current component I_5, the ratio of input current to the phase current at full load is 1.92. This is higher than the ratio of $\sqrt{3}$ in the case of a single-phase IG with the Steinmetz connection, and may be attributed to the primary current I_{51} of the current injection transformer.

Figure 3.48 shows the efficiency and power factor characteristics of the IG under perfect phase balance. As the rotor speed varies from 1533 to 1570 r/min, the efficiency decreases almost linearly from 0.74 to 0.70 p.u. The output power factor, on the other hand, increases from 0.85 leading to unity over the same speed range. Above 1570 r/min, the power factor decreases slightly and becomes lagging. This

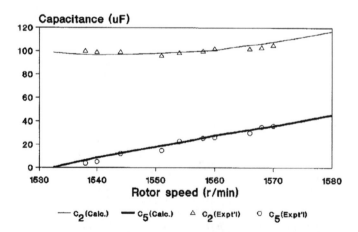

Figure 3.46 *Capacitances to give perfect phase balance at different speeds*

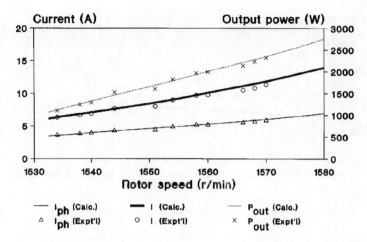

Figure 3.47   *Phase current, line current and output power at perfect phase balance*

novel phase balancing method therefore results in excellent operating line power factors for the experimental machine.

The close agreement between the computed and experimental results in Figures 3.46 to 3.48 confirms the principle of phase balancing as explained in Section 3.5.1.

A load test was conducted on the experimental machine with $C_2 = 102\mu F$ and $C_5 = 30\mu F$. It was found that the IG was balanced at a rotor speed of 1566 r/min and at a phase current of 5.5 A. Figures 3.49–3.52 show the computed and experimental performance characteristics obtained. As shown in Figure 3.49, the B-phase voltage V_B increases as the rotor speed decreases from that corresponding to the balance point, whereas the C-phase voltage V_C decreases. Figure 3.50 shows that overcurrent occurs in B-phase at rotor speeds below 1550 r/min. I_A and I_C, on the

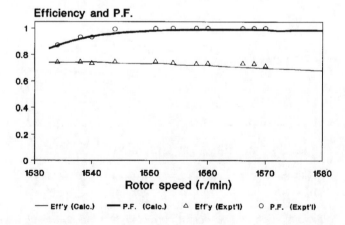

Figure 3.48   *Efficiency and output power of IG at perfect phase balance*

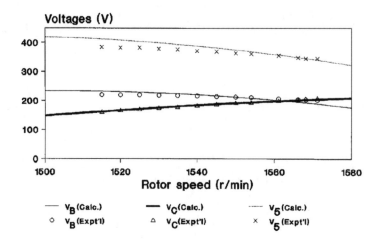

Figure 3.49   *Variations of voltages V_B, V_C and V_5 when the IG operates with constant capacitances*

other hand, are less than the rated value and hence offset the effect of the overcurrent in B-phase. At a rotor speed of 1510 r/min, which corresponds to a 'floating' condition, the total stator copper losses are still within permissible limits.

Figure 3.51 and Figure 3.52 show that, when the capacitances are constant, the efficiency, the power factor and the output power are in general lower than when the capacitances are varied to give balanced operation. A comparison between Figure 3.47 and Figure 3.52 also shows that the output power is lower when the IG operates with constant capacitances. This is due to the negative-sequence losses under unbalanced operation.

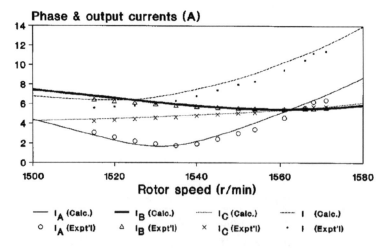

Figure 3.50   *Variations of phase and output currents when the IG operates with constant capacitances*

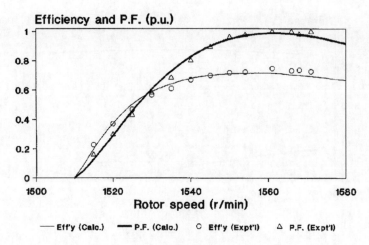

Figure 3.51 *Variations of efficiency and power factor when the IG operates with constant capacitances*

Very good agreement between the computed and experimental results is observed in Figures. 3.49–3.52, thereby confirming the validity of the method of analysis.

Figure 3.53 shows the computed variations of the positive-sequence voltage V_p and negative-sequence voltage V_n when $C_2 = 102\mu F$ and $C_5 = 30\mu F$. V_p exhibits a slightly convex downward characteristic: the maximum value occurs at a rotor speed of 1545 r/min, but over the normal speed range the variation in V_p is less than 5%. The variation of V_n, on the other hand, is in the form of a V-shaped curve, with zero value occurring at a rotor speed of 1562.5 r/min. The variation of the VUF (which is the ratio of magnitudes of V_p and V_n) is thus also a V-shaped curve.

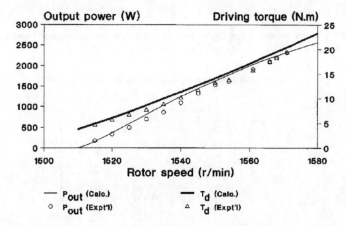

Figure 3.52 *Variations of output power and driving torque when the IG operates with constant capacitances*

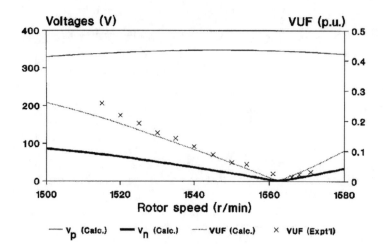

Figure 3.53   *Variations of V_p, V_n and VUF when the IG operates with constant capacitances*

Waveforms and harmonics. Figures 3.54–3.56 show the waveforms of the experimental IG system when operating under perfect phase balance with a phase
current of 5.0 A, measured using a dual-channel digital storage oscilloscope. In
each oscillogram, the A-phase voltage (upper trace) was measured together with
the waveform of the quantity of interest (lower trace). The phase angle between

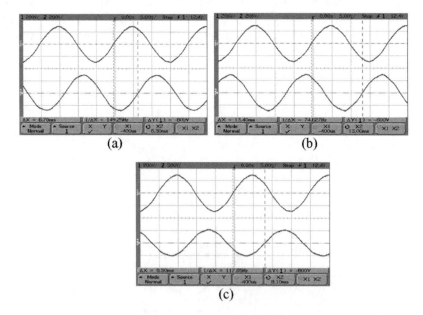

Figure 3.54   *Voltage waveforms of experimental IG at full load: (a) V_A and V_B; (b) V_A and V_C;
(c) V_A and V_5 (scale for V_A, V_B and V_C: 200 V/div; scale for V_5: 500 V/div; time scale: 5 ms/div)*

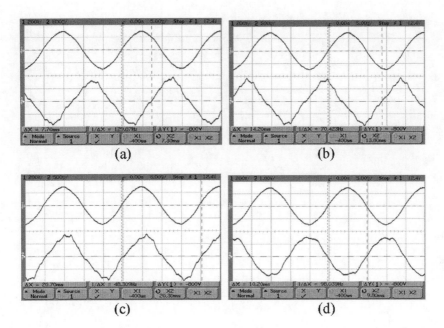

(a) (b)

(c) (d)

Figure 3.55 *Phase and line current waveforms of experimental IG at full load: (a) V_A and I_A; (b) V_A and I_B; (c) V_A and I_C; (d) V_A and I (scale for phase currents: 5 A/div; scale for line current: 10 A/div; time scale: 5 ms/div)*

the upper and lower waveforms is given by the separation of the vertical cursors X1 and X2 which were set to the zero crossings of the two waveforms.

The waveforms in Figures 3.54(a) and (b) show that the phase voltages of the IG are quite sinusoidal. A close examination of the phase angles of the waveforms confirmed that the phase voltages were balanced. The third-harmonic induced e.m.f.s, however, cause a circulating current to flow in the delta-connected stator winding. This can be observed from the phase current waveforms shown in Figures 3.55(a) to (c). Except for the third-harmonic distortion, the phase currents are balanced and they have approximately the same wave shape. From Figure 3.55(d), it is observed that the line current I and V_A are almost in anti-phase, which implies that the output power factor is very close to unity.

Figures 3.56(a) and (b) show the waveforms of the capacitor currents I_2 and I_5 while Figure 3.56(c) shows the waveform of I_{51}. Harmonic distortion is more prominent in these waveforms since a capacitance presents lower impedance to higher order current harmonics.

The principal voltage and current harmonics of the IG were measured using a harmonic analyser. It was found that in all the system waveforms, harmonics of the nineth order and above were negligible. Tables 3.8 and 3.9 show the magnitudes of the principal harmonics in the voltages and currents. The results in Table 3.8 confirm the sinusoidal waveforms obtained in Figure 3.54. It is seen that third-harmonic distortion in the phase voltages is quite small, but the seventh harmonic

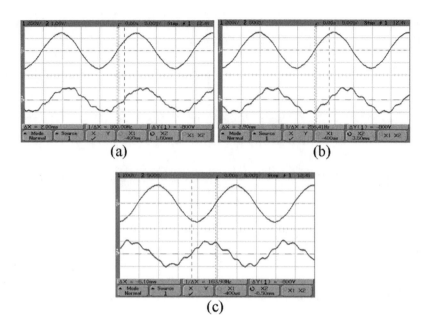

Figure 3.56 *Waveforms of capacitor currents and transformer primary current: (a) V_A and I_2; (b) V_A and I_5; (c) V_A and I_{51} (scale for I_2: 10 A/div; scale for I_5 and I_{51}: 5 A/div; time scale: 5 ms/div)*

is more prominent. The results in Table 3.9, on the other hand, show that the harmonic distortion in the phase currents is mainly due to the third-harmonic, which is consistent with the observed waveforms in Figure 3.55. The capacitor currents I_2 and I_5 are rich in the seventh harmonic. In the case of I_5, the higher seventh-harmonic content may be due to the smaller effective capacitive reactance since C_5 is in series with the transformer equivalent leakage reactance. The phase currents of the IG, on the other hand, contains very little of the seventh harmonic. The primary current I_{51} of the current injection transformer has the most severe harmonic distortion due to the seventh harmonic reflected from the secondary winding and the third-harmonic component in the no-load current. In the experimental IG system, however, the magnitude of I_{51} is small compared with the line current. The line

Table 3.8 *Principal voltage harmonics in experimental IG when balanced at a phase current of 5 A (expressed as percentage of the fundamental)*

Harmonic order	V_A	V_B	V_C	V_5
3	0.29	1.0	0.84	1.03
5	1.17	0.55	1.51	0.73
7	2.03	1.98	3.72	3.76

Table 3.9   *Principal current harmonics in experimental IG when balanced at a phase current of 5 A (expressed as percentage of the fundamental)*

Harmonic order	I_A	I_B	I_C	I	I_2	I_5	I_{51}
3	8.63	10.84	9.7	4.42	3.34	2.2	7.1
5	1.77	1.33	2.8	2.79	2.68	3.07	1.9
7	1.55	1.77	0.0	5.35	12.9	25.4	29.0

current waveform is therefore quite sinusoidal, with a total harmonic distortion (THD) of 7.5 % approximately.

3.5.5   Summary

This section has presented the principle and performance analysis of Smith's Mode C circuit that enables a three-phase IG to be operated satisfactorily on a single-phase grid. The circuit employs a transformer and two capacitances for phase balancing. A phasor diagram approach enables analytic expressions for the values of capacitances to be deduced. A general performance analysis is developed by applying the method of symmetrical components. Experimental results obtained on a small induction machine have confirmed the accuracy of analysis and feasibility of the circuit configuration. The advantageous features of this phase balancing scheme include simplicity of the circuit configuration, low cost, high efficiency and excellent system power factor.

References

[1] P.G. Holmes, 'Single- to 3-phase transient phase conversion in induction motor drives,' *IEE Proceedings*, Pt B, Vol. 132, No. 5, pp. 289–296, September 1985.

[2] T.F. Chan, 'Performance analysis of a three-phase induction generator connected to a single-phase power system,' *IEEE Transactions on Energy Conversion*, Vol. 13, No. 3, pp. 205–211, September 1998.

[3] C.E. Tindall and W. Monteith, 'Balanced operation of 3-phase induction motors connected to single-phase supplies,' *IEE Proceedings*, Vol. 123, No. 6, pp. 517–522, June 1976.

[4] J.E. Brown and O.I. Butler, 'A general method of analysis of 3-phase induction motors with asymmetrical primary connections,' *IEE Proceedings*, Pt II, Vol. 100, pp. 25–34, February 1953.

[5] V. Ostovic, *Dynamics of Saturated Electric Machines*, Springer-Verlag, New York, 1989.

[6] T.F. Chan and L.L. Lai, 'Steady-state analysis of a three-phase induction motor with the Smith connection,' *IEEE Power Engineering Review*, Vol. 20, No. 10, pp. 45–46, October 2000.

[7] T.F. Chan and L.L. Lai, 'Single-phase operation of a three-phase induction motor with the Smith connection,' *Proceedings of 2000 IEEE Power Engineering Society Summer Meeting*, 18–20 July 2000, Seattle, WA, USA, Vol. 4, pp. 2453-2458.

[8] L.S. MacKenzie, *The 8051 Microcontroller*, Prentice Hall, Englewood Cliffs, NJ, 1995.

[9] S. Yeralan and A. Ahluwalia, *Programming and Interfacing the 8051 Microcontroller*, Addison-Wesley, Reading, MA, 1995.

[10] O.J.M. Smith, 'High-efficiency single-phase motor,' *IEEE Transactions on Energy Conversion*, Vol. 7, No. 3, pp. 560–569, September 1992.

[11] O.J.M. Smith, 'Three-phase induction generator for single-phase line,' *IEEE Transactions on Energy Conversion*, Vol. EC-2, No. 3, pp. 382–387, September 1987.

4

Finite Element Analysis of Grid-Connected IG with the Steinmetz Connection

4.1 Introduction

Among the various phase balancing schemes in use, the Steinmetz connection [1] is particularly suitable for adapting a three-phase induction machine for single-phase operation due to the extremely simple circuit configuration. The method is applicable to an induction machine with a star-connected winding or a delta-connected stator winding. It can be shown that, for a given value of phase balancing capacitance, there exists a speed or load at which the voltage unbalance factor is a minimum [2]. Furthermore, if the values of the capacitance and speed are properly chosen, a perfect phase balance condition can be achieved. With the aid of the Steinmetz connection, three-phase induction machines could conveniently be used as non-utility generators in distributed single-phase generation systems in remote or rural regions.

In Chapter 3, the performance analysis of an IG with asymmetrical winding connection or unbalanced phase voltages is carried out using the method of symmetrical components. Since the method is based on the principle of superposition, the assumption of circuit linearity has to be made. When the induction machine operates in the generator mode, the higher air gap voltage leads to more severe magnetic saturation, and consequently the assumption of a linear magnetic circuit is less valid. The accuracy of the symmetrical components method is also affected by the variation in the values of the equivalent circuit parameters, in particular the rotor resistance which depends on the rotor slip and thermal conditions. In order to account for the complex magnetic field distribution in the IG and the dependence of

Distributed Generation: Induction and Permanent Magnet Generators L. L. Lai and T. F. Chan
© 2007 John Wiley & Sons, Ltd

rotor parameters on speed, detailed modelling and solution of the machine variables in the magnetic field domain will be more appropriate.

Electromagnetic field analysis of three-phase induction motor drives using the finite element method (FEM) has been a subject of rigorous research in the past two decades (see Appendix C) [3–8]. The major advantage of the FEM is the accurate modelling of the machine's magnetic circuit. Magnetic nonlinearity, material non-homogeneity and the effect of discrete winding layouts can easily be accounted for. Another advantage of the FEM is that it permits direct modelling in phase quantities, hence the analysis implicitly includes the effects of space and time harmonics. The third advantage of the FEM is its ability to deal with both steady-state and transient machine operations, without requiring simplifying assumptions such as those made in classical machine theory, e.g. the d–q machine model as used in the Electromagnetic Transients Program (EMTP) [9,10]. The machine performance can be obtained directly from the field solution, without the need to use the equivalent circuit parameters. A feature of the FEM, however, is the relatively long computing time and the large memory requirement.

Most of the publications on the FEM have been devoted to three-phase induction motor drives under balanced operating conditions [3, 5, 6]. To the author's best knowledge, FEM analysis of a three-phase IG with the Steinmetz connection has not been reported before.

In this chapter, the performance analysis of a single-phase grid-connected IG with the Steinmetz connection will be conducted using a coupled circuit and field approach, based on a two-dimensional (2-D) FEM. Emphasis will be placed on the phase balancing capability of the IG circuit configuration under steady-state conditions. The generator performance computed by the proposed method will be compared with that computed by the method of symmetrical components and that obtained from experiments on a 2.2 kW induction machine.

4.2   Steinmetz Connection and Symmetrical Components Analysis

The analysis of the Steinmetz connection shown in Fig. 4.1 is quite straightforward and involves the following steps:

1. Establish the 'inspection equations' based on Kirchhoff's laws.
2. Solve the 'inspection equations' in association with the symmetrical component's equations to obtain the positive-sequence voltage V_p and negative-sequence voltage V_n.
3. Compute the positive-sequence current I_p and negative-sequence current I_n from the equivalent circuits of the IG.
4. Compute the performance of the IG.

It is well known that the output power factor of an IG is leading, which implies that the air gap voltage is considerably higher than the corresponding value when the

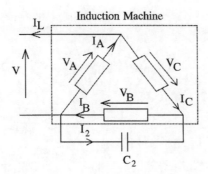

Figure 4.1 *Single-phase operation of three-phase, delta-connected IG with Steinmetz connection. Reproduced by permission of T.-F. Chan, L. L. Lai and L.-T. Yan, 'Finite element analysis of a single-phase grid-connected induction generator with the Steinmetz connection'.* IEEE *Transactions on Energy Conversion, 18, 2003; 321–329. © (2003) IEEE*

machine operates as a motor. Magnetic saturation is more severe and hence it is very difficult to choose the proper value of magnetizing reactance that truly reflects the magnetic saturation conditions in the machine. The magnetic field distribution is further complicated by the presence of the negative-sequence rotating field, which in turn depends upon the rotor speed or output power. The deep-bar effect in the rotor conductors also causes a discrepancy between the effective value of rotor resistance and that determined from standard locked-rotor and DC resistance tests. Due to the above uncertainties, computations based on the method of symmetrical components are subject to errors. In order to account for the complex magnetic conditions in the machine, a field approach will be attempted as detailed in the following sections.

4.3 Machine Model

A coupled circuit and field approach is adopted for performance analysis of a three-phase IG with the Steinmetz connection. Figure 4.2 shows a cross-sectional view of the experimental machine chosen for the investigation. Pertinent technical data of the induction machine are given in Appendix D.2. To reduce the complexity of the problem and solution time, the following assumptions and solution techniques will be used:

1. The 2-D FEM is used for the magnetic field computation, with the complex magnetic vector potentials as the variables. Rectangular $(x-y)$ coordinates are used in the FEM model, hence each magnetic vector potential A has only an axial or z-component [4].
2. First-order triangular elements and linear interpolation functions are used.
3. Stator end-winding leakage reactance is ignored.

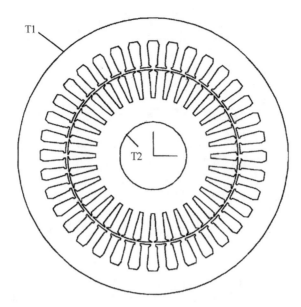

Figure 4.2   *Cross-section of experimental IG. Reproduced by permission of T.-F. Chan, L. L. Lai and L.-T. Yan, 'Finite element analysis of a single-phase grid-connected induction generator with the Steinmetz connection'. IEEE* Transactions on Energy Conversion, *18, 2003; 321–329.* © *(2003) IEEE*

4. The rotor speed is constant.
5. The nominal magnetization curve is used for both stator and rotor iron cores, i.e. the hysteresis effect is neglected.

4.4   Finite Element Analysis

Since the stator winding connection is asymmetrical, the electromagnetic quantities such as voltage, current and flux density are all time varying but may not be sinusoidal. Accordingly, it is not possible to use the complex form of the field equation. A time-stepping finite element field formulation coupled with external circuit equations is therefore adopted in the solver. Besides, the solution region has to be extended to the entire cross-section of the induction machine.

4.4.1   Basic Field Equations

Using the assumptions made in Section 4.3 and with reference to Figure 4.2, the basic field equation and boundary conditions are formulated as follows [11, 12]:

$$\frac{\partial}{\partial x}\left(\frac{1}{\mu}\cdot\frac{\partial A}{\partial x}\right) + \frac{\partial}{\partial y}\left(\frac{1}{\mu}\cdot\frac{\partial A}{\partial y}\right) = -J \qquad (4.1)$$

$$A\mid_{T1} = A\mid_{T2} = 0 \qquad (4.2)$$

where $A =$ magnetic vector potential;
$J =$ externally impressed current density;
$\mu =$ magnetic permeability;
$T1 =$ outer periphery of the stator iron core;
$T2 =$ inner periphery of the rotor iron core.

The energy functional is

$$E(A) = \iint_{\Omega} \left(\int_0^B \frac{B}{\mu}.dB - A.J \right) dx.dy \tag{4.3}$$

where Ω denotes the field solution region and B is the flux density.

After discretization and functional minimization, the following matrix equation is obtained:

$$[K][A] = [R] \tag{4.4}$$

where $[K] =$ coefficient matrix;
$[A] =$ column vector of nodal magnetic vector potentials;
$[R] =$ right-hand-side column matrix containing known terms.

Equation (4.4) needs to be modified due to the coupled circuit and field approach used in the solution. Details of the circuit formulation will be given in the following sections.

4.4.2 Stator Circuit Equations

Figure 4.3 shows the stator circuit model for single-phase operation of the IG with the Steinmetz connection. E_A, E_B and E_C are the internal e.m.f.s induced in stator phases A, B and C, respectively. The generator convention will be adopted in the formulation of the circuit equations.

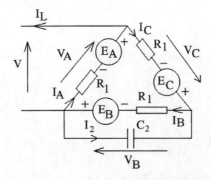

Figure 4.3 *Stator circuit model for delta-connected IG with Steinmetz connection. Reproduced by permission of T.-F. Chan, L. L. Lai and L.-T. Yan, 'Finite element analysis of a single-phase grid-connected induction generator with the Steinmetz connection'. IEEE* Transactions on Energy Conversion, *18, 2003; 321–329. © (2003) IEEE*

Neglecting the stator end-winding leakage reactance, the following equation may be written for A-phase:

$$V_A = E_A - I_A R_1.$$ (4.5)

To facilitate easy coupling with the FEM field domain, (4.5) is rewritten as

$$I_A - \frac{E_A}{R_1} = -\frac{V_A}{R_1} = -\frac{V}{R_1}.$$ (4.6)

Since the impressed voltages across B-phase and C-phase are not explicitly known, they have to be expressed in terms of other circuit parameters, available after the computation for the previous time step. For B-phase,

$$V_B = E_B - I_B R_1.$$ (4.7)

The B-phase and C-phase currents are related by

$$I_C + I_2 = I_B.$$ (4.8)

The capacitor current I_2 can be written as

$$I_2 = C_2 \frac{dV_B}{dt} \approx C_2 \cdot \frac{V_B - V_B'}{\Delta t}$$ (4.9)

where $V_B\prime$ = value of B-phase voltage in the previous time step;
Δt = incremental time step.

Substituting (4.8) and (4.9) into (4.7) and simplifying, the following equation may be written:

$$\left(1 + \frac{\Delta t}{R_1 C_2}\right) I_B - \frac{\Delta t}{R_1 C_2} I_C - \frac{E_B}{R_1} = \frac{V_B'}{R_1}.$$ (4.10)

For C-phase,

$$V_C = E_C - I_C R_1.$$ (4.11)

The sum of the phase voltages must be equal to zero for a delta circuit, hence

$$V_A + V_B + V_C = 0.$$ (4.12)

Eliminating V_C from (4.11) and (4.12),

$$-V_A - V_B = E_C - I_C R_1.$$ (4.13)

Eliminating V_B from (4.9) and (4.13),

$$V_A + \frac{\Delta t}{C_2} I_2 + V_B' + E_C - I_C R_1 = 0.$$ (4.14)

Finally, eliminating I_2 from (4.8) and (4.14),

$$-\frac{\Delta t}{R_1 C_2} I_B + \left(1 + \frac{\Delta t}{R_1 C_2}\right) I_C - \frac{E_C}{R_1} = \frac{V_A + V_B'}{R_1}.$$ (4.15)

It should be noted that (4.6), (4.10) and (4.15) have been expressed in a form suitable for coupling with the matrix equation (4.4). All the additional circuit variables are grouped on the left-hand side of each equation, while the known terms are placed on the right-hand side.

4.4.3 Stator EMFs

From the FEM solution, the average value of magnetic vector potential in an element e is

$$A_e - \frac{1}{S} \int\int_\Delta A.dx.dy -- \frac{1}{S} \int\int N_t \cdot A_t \cdot dx.dy = \frac{1}{S} \sum_{i=1}^{3} \frac{\Delta e}{3} \cdot A_t \qquad (4.16)$$

where N_i = shape function;
S = cross-sectional area of coil side;
Δe = area of triangular element.

The stator internal e.m.f. E_{ph} (ph = A, B and C) can be computed next by applying Faraday's law, using the fact that flux density is the time derivative of the magnetic vector potential A:

$$E_{ph} = -\sum_{e=1}^{n} \left(-\frac{\partial}{\partial t} A_e \right) .l_{fe}.\frac{N_c}{2}.k_e = -\sum_{e=1}^{n} \left(\frac{\Delta A_e}{\Delta t} \right) .l_{fe}.\frac{N_c}{2}.k_e$$
$$= \frac{N_c.l_{fe}}{2\Delta t} \left(\sum_{e=1}^{n} k_e A_e - \sum_{e=1}^{n} k_e A'_e \right) \qquad (4.17)$$

where A'_e = average value of vector potential of element computed in the previous time step;
N_c = turns per coil;
l_{fe} = axial length of iron core;
k_e = ±1, depending on the position of the coil side (i.e. whether it is at the starting end or finishing end of the coil).

4.4.4 Rotor Circuit Model

Under normal operating conditions, the frequency of the positive-sequence rotor e.m.f. and current is very low, typically 2 to 3Hz. This implies that the effective rotor resistance is much smaller than that deduced from a standard locked-rotor test conducted at rated frequency. On the contrary, the negative-sequence rotor current is approximately equal to twice the rated frequency, hence the effective negative-sequence rotor resistance is higher than the locked-rotor value. Due to the uncertainties in the value of rotor resistance, solution using the method of symmetrical components is subject to appreciable error.

Rotor end-winding leakage reactance is neglected in some FEM programs for balanced operation of a symmetrical three-phase induction machine. Such an

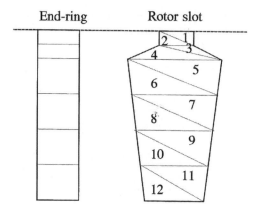

Figure 4.4 *Shape of rotor conductor and end ring: each rotor conductor is partitioned into 12 elements in six layers for field computation. Reproduced by permission of T.-F. Chan, L. L. Lai and L.-T. Yan, 'Finite element analysis of a single-phase grid-connected induction generator with the Steinmetz connection'. IEEE* Transactions on Energy Conversion, *18, 2003; 321–329.* © *(2003) IEEE*

assumption, however, cannot be used for the present problem as this would lead to very large pulsation in the computed rotor current and non-convergence.

A rotor circuit model is therefore introduced to overcome the above difficulties. Figure 4.4 shows the sectional view of the rotor conductor bar and end ring of the experimental machine. To account for the frequency effect, each rotor bar is partitioned into six layers, each with two triangular elements. The current density in each layer is assumed to be constant during the field solution. Each end ring is likewise partitioned into six layers. For each layer, the resistance of the rotor conductor R_{21}, the cross-sectional area of rotor conductor A_{RL}, the resistance of the end ring R_{22}, and the leakage inductance of the end ring L_2, can be computed. Figure 4.5 shows the equivalent circuit of each layer of the rotor cage winding, where $J3, J4, J5, \ldots$ denote the corresponding layers in successive rotor slots.

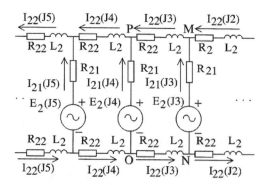

Figure 4.5 *Circuit model of rotor cage winding. Reproduced by permission of T.-F. Chan, L. L. Lai and L.-T. Yan, 'Finite element analysis of a single-phase grid-connected induction generator with the Steinmetz connection'. IEEE* Transactions on Energy Conversion, *18, 2003; 321–329.* © *(2003) IEEE*

Consider a typical rotor conductor layer MN in Figure 4.5, i.e. the one with generated e.m.f. $E_2(J3)$. Three new circuit variables are introduced, namely the current density in the layer $\delta(J3)$, end-winding current $I_{22}(J3)$ and conductor e.m.f. $E_2(J3)$.

Applying Kirchhoff's current law to node M,

$$I_{22}(J2) + I_{21}(J3) = I_{22}(J3). \tag{4.18}$$

The rotor bar layer current $I_{21}(J3)$ can be expressed as

$$I_{21} = \delta(J3). \, A_{RL} . \tag{4.19}$$

Applying Kirchhoff's voltage law to mesh MNOP in Figure 4.5,

$$E_2(J3) - E_2(J4) = I_{21}(J3). \, R_{21} + 2 \, I_{22}(J3). \, R_{22}$$
$$+ 2 \, L_2 . \frac{d}{dt} \, I_{22}(J3) - I_{21}(J4). \, R_{21} . \tag{4.20}$$

Rewriting (4.20) as a difference equation,

$$E_2(J3) - E_2(J4) - \delta(J3). \, A_{RL} . \, R_{21} - 2 \, I_{22}(J3). \, R_{22}$$
$$- 2 \, L_2 . \, I_{22}(J3)/\Delta t + \delta(J4). \, A_{RL} . \, R_{21} = -2 \, L_2 . \, I_{22}'(J3)/\Delta t. \tag{4.21}$$

In (4.21), $I_{22}'(J3)$ is the value of $I_{22}(J3)$ in the previous time step and becomes a known quantity for the current FEM computation.

The rotor e.m.f. $E_2(J3)$ is taken to be the average value of the induced e.m.f. in the two elements (say e and $e + 1$) in a rotor conductor layer, i.e.

$$E_2(J3) = \frac{E(e) + E(e + 1)}{2}. \tag{4.22}$$

For element e, the e.m.f. $E(e)$ is the sum of transformer e.m.f. and motional e.m.f., i.e.

$$E(e) = -\frac{\partial A}{\partial t} . \, l_{fe} + \overline{v} \times \overline{B} . \, l_{fe} . \tag{4.23}$$

To evaluate the first term in (4.23), the derivative may be approximated by

$$\frac{\partial A}{\partial t} = \frac{A_e - A_e'}{\Delta t} \tag{4.24}$$

where A_e = average value of magnetic vector potential of element e at time t;
A_e' = average value of magnetic vector potential of element e at time $t - \Delta t$.

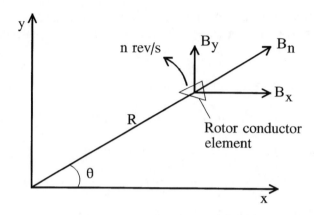

Figure 4.6 *Computation of flux density and generated e.m.f. in rotor conductor element. Reproduced by permission of T.-F. Chan, L. L. Lai and L.-T. Yan, 'Finite element analysis of a single-phase grid-connected induction generator with the Steinmetz connection'. IEEE Transactions on Energy Conversion,* **18**, *2003; 321–329.* © *(2003) IEEE*

Figure 4.6 shows an arbitrary conductor element at radius R from the axis of rotation. If the rotor rotates at a speed n r/s, the linear velocity of the element is

$$v = 2\pi r.n. \tag{4.25}$$

The components of flux density at the rotor element are

$$B_x = \frac{\partial A}{\partial y} = \frac{CL(1).\, A_i + CL(2).\, A_j + CL(3).\, A_k}{2\Delta} \tag{4.26}$$

$$B_y = -\frac{\partial A}{\partial x} = \frac{-BL(1).\, A_i - BL(2).\, A_j - BL(3).\, A_k}{2\Delta} \tag{4.27}$$

where $CL(\kappa)$ and $BL(\kappa)$ ($\kappa = 1, 2, 3$) are constants pertaining to the element being considered and Δ is the area of the element.

The component of flux density normal to the linear velocity of the element is

$$B_n = B_x.\cos\theta + B_y.\sin\theta = \gamma_i.\, A_i + \gamma_j.\, A_j + \gamma_k.\, A_k \tag{4.28}$$

where

$$\begin{aligned}
\gamma_i &= CL(1).\cos\theta - BL(1).\sin\theta;\\
\gamma_j &= CL(2).\cos\theta - BL(2).\sin\theta;\\
\gamma_k &= CL(3).\cos\theta - BL(3).\sin\theta.
\end{aligned} \tag{4.29}$$

4.4.5   Comments on the Proposed Method

A time-stepping FEM coupled with external circuit equations for the performance analysis of an IG with the Steinmetz connection has been developed. The capacitance in the stator circuit and end-winding leakage reactance in the rotor circuit both involve differential equations, and hence a time-varying transient circuit model is adopted. In this respect, the proposed approach is not fundamentally different from the time-stepping coupled finite element–state space (TSCFE-SS) algorithm [3, 7, 8] used by other researchers. In the TSCFE-SS method, inductance profiles have to be generated for each time step in order to determine the excitation currents for the subsequent time step. The proposed method, however, aims to compute the machine performance directly from the field solution. Another salient feature of the proposed method is the use of a refined rotor circuit model that enables the complex current distribution in the rotor circuit to be considered.

In the present IG configuration, the grid voltage (i.e. V_A) is known and is taken to be a time-varying sinusoidal quantity. For the other phase voltages and currents, the magnitudes and the waveforms are obtained from the FEM solver and hence the effects of time and space harmonics on the generator performance are automatically accounted for.

The present FEM solver could be adapted for dynamic performance studies, such as turbine speed changes and switching operations. This requires coupling of the electromechanical equation [6] with the matrix equations developed earlier in this chapter and the rotor speed will then be one of the machine variables.

As far as possible the proposed method avoids the use of induction machine equivalent circuit parameters, defined conventionally under sinusoidal current and voltage conditions. The only assumption made is the omission of the stator end-winding leakage reactance, which simplifies the stator circuit equations with only a slight loss in accuracy.

4.5   Computational Aspects

A FORTRAN computer program was developed for the performance analysis of the three-phase IG with the Steinmetz connection. The computations referred to the experimental machine IG2 whose technical data are given in Appendix D.2. Program runs were conducted on a Pentium 586 computer with a clock speed of 333MHz, 128MB of RAM and 8GB hard disk. In view of the constraints on the computing facilities, a compromise had to be made among the computational accuracy, run time and the memory requirements. Figure 4.7 shows the finite element mesh used for the field analysis, the number of nodes and elements being equal to 2120 and 4068, respectively. Since 172 nodes fall on the boundaries with constant values of magnetic vector potential, the number of nodal magnetic vector potentials to be evaluated is 1948. Table 4.1 shows the number of variables used in the FEM solver.

The time for solution of the matrix equation was approximately 45 seconds. For the electromagnetic quantities to settle to a steady state, the computation had to be extended to 10 complete electrical cycles and the number of time steps (each corresponding to one electrical degree) required was over 1500. The total time for computing an operating point therefore amounted to 7 to 8 hours.

Figure 4.7   *Finite element mesh for experimental machine. Reproduced by permission of T.-F. Chan, L. L. Lai and L.-T. Yan, 'Finite element analysis of a single-phase grid-connected induction generator with the Steinmetz connection'. IEEE* Transactions on Energy Conversion, **18**, *2003; 321–329.* © *(2003) IEEE*

4.6   Case Study

Laboratory tests were conducted on the delta-connected experimental machine referred to in the previous section. The IG was operated on a 220 V, 50 Hz single-phase supply. A phase converter capacitance of 93.6 μF was used to give approximately

Table 4.1   *Distribution Of Variables In FEM Program. Reproduced by permission of T.-F. Chan, L. L. Lai and L.-T. Yan, 'Finite element analysis of a single-phase grid-connected induction generator with the Steinmetz connection'. IEEE* Transactions on Energy Conversion, **18**, *2003; 321–329.* © *(2003) IEEE*

Circuit/field variables	Number
Nodal magnetic vector potentials, A	1948
Stator phase e.m.f.s	3
Stator phase currents	3
Rotor circuit variables	576
Total	2530

balanced operation at a speed of 1540 r/min and a phase current of 4.0 A. Instruments were arranged to measure the speed, phase currents, phase voltages, line current and output power. To compare the performance of the IG obtained by the method of symmetrical components, a standard no-load test, a locked-rotor test and a DC resistance test were conducted separately, from which the following parameters were determined (assuming that the stator and referred rotor leakage reactances are equal):

Stator leakage impedance	$= (3.08 + j5.68)\Omega$
Rotor leakage impedance referred to the stator	$= (2.85 + j5.68)\Omega$
Magnetizing reactance	$= 80.8\Omega$
Core loss resistance	$= 1463\Omega$
Friction and windage loss	$= 27W.$

Figures 4.8–4.14 show the computed and experimental results obtained on the IG with the Steinmetz connection. From Figure 4.8 and Figure 4.9, it is observed that the method of symmetrical components gives an accurate prediction of the variation of V_B and V_C. In comparison, the FEM gives good prediction of V_C but rather poor prediction of V_B, especially at low speeds.

From Figures 4.10 to 4.12, it is found that both the FEM and method of symmetrical components give correct prediction of the trend of the phase currents. At high speeds, the FEM gives a more accurate prediction of I_A and I_C, whereas the method of symmetrical components gives a better prediction of I_B.

Figure 4.13 and Figure 4.14 show the variation of line current and output power of the IG with rotor speed. Over the entire operating speed range, the FEM gives more accurate results compared with the method of symmetrical components.

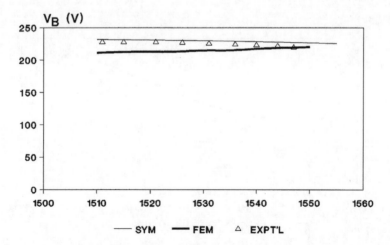

Figure 4.8 *Variation of B-phase voltage with rotor speed. Reproduced by permission of T.-F. Chan, L. L. Lai and L.-T. Yan, 'Finite element analysis of a single-phase grid-connected induction generator with the Steinmetz Connection'. IEEE Transactions on Energy Conversion, 18,2003;321–329. © (2003) IEE*

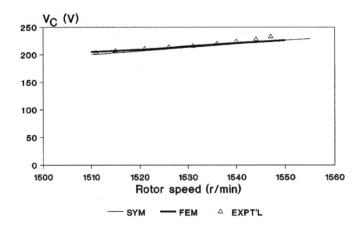

Figure 4.9   *Variation of C-phase voltage with rotor speed. Reproduced by permission of T.-F. Chan, L. L. Lai and L.-T. Yan, 'Finite element analysis of a single-phase grid-connected induction generator with the Steinmetz connection'. IEEE* Transactions on Energy Conversion, **18***, 2003; 321–329.* © *(2003) IEEE*

Table 4.2 summarizes the deviations in generator performance computed by the method of symmetrical components and the FEM. For easy comparison, the deviation for each machine variable has been normalized to the respective experimental value. It is observed that both methods yield large deviations at low speed (1511r/min). At high speed (1547r/min), the FEM gives more accurate results in all the variables except the B-phase current. The ability of the FEM to account for magnetic saturation and change in effective rotor resistance due to loading and circuit imbalance is thus confirmed.

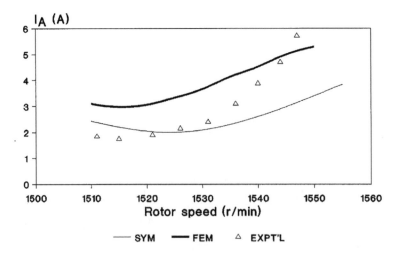

Figure 4.10   *Variation of A-phase current with rotor speed. Reproduced by permission of T.-F. Chan, L. L. Lai and L.-T. Yan, 'Finite element analysis of a single-phase grid-connected induction generator with the Steinmetz connection'. IEEE* Transactions on Energy Conversion, **18***, 2003; 321–329.* © *(2003) IEEE*

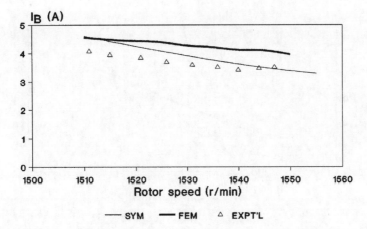

Figure 4.11 *Variation of B-phase current with rotor speed. Reproduced by permission of T.-F. Chan, L. L. Lai and L.-T. Yan, 'Finite element analysis of a single-phase grid-connected induction generator with the Steinmetz connection'. IEEE Transactions on Energy Conversion,* **18**, *2003; 321–329.* © *(2003) IEEE*

The above results indicate that discrepancies exist between the computed and experimental performance of the IG with the Steinmetz connection. In the case of the method of symmetrical components, the deviations mainly exist in the currents and the output power, which is due to the limitations in the circuit model. In the case of the FEM, large deviations occur in the B-phase voltage, A-phase current and B-phase current. These discrepancies may be attributed to the following factors:

1. Numerical solution of the circuit differential equations requires their transformation to the corresponding difference equations, which involves the time step Δt. A smaller Δt reduces the discretization error. At the same time more air

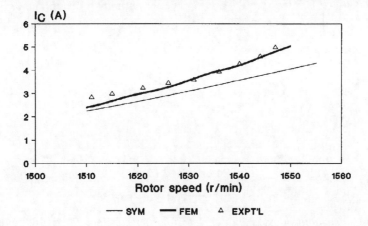

Figure 4.12 *Variation of C-phase current with rotor speed. Reproduced by permission of T.-F. Chan, L. L. Lai and L.-T. Yan, 'Finite element analysis of a single-phase grid-connected induction generator with the Steinmetz connection'. IEEE Transactions on Energy Conversion,* **18**, *2003; 321–329.* © *(2003) IEEE*

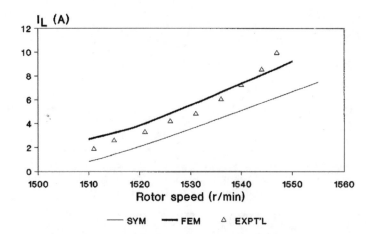

Figure 4.13 *Variation of line current with rotor speed. Reproduced by permission of T.-F. Chan, L. L. Lai and L.-T. Yan, 'Finite element analysis of a single-phase grid-connected induction generator with the Steinmetz connection'. IEEE Transactions on Energy Conversion, 18, 2003; 321–329.* © *(2003) IEEE*

gap nodes must be used, which is beneficial in reducing the error caused by distortion of the air gap elements as a result of the rotor rotation (time stepping).
2. Numerical solution of a high-order matrix equation incurs considerable cumulative errors. This problem may be resolved by using double precision in the FEM solver, but the memory requirement will be increased significantly.
3. Thermal effects on the machine parameters, particularly on the rotor resistance, are difficult to account for. In the FEM computations, the rotor resistance was

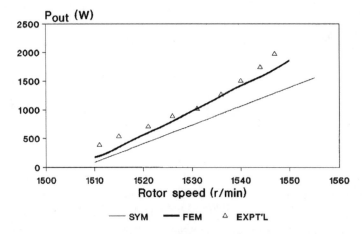

Figure 4.14 *Variation of output power with rotor speed. Reproduced by permission of T.-F. Chan, L. L. Lai and L.-T. Yan, 'Finite element analysis of a single-phase grid-connected induction generator with the Steinmetz connection'. IEEE Transactions on Energy Conversion, 18, 2003; 321–329.* © *(2003) IEEE*

Table 4.2 *Deviations in IG performance computed by the method of symmetrical components (SYM) and the finite element method (FEM) (normalized to respective experimental values). Reproduced by permission of T.-F. Chan, L. L. Lai and L.-T. Yan, 'Finite element analysis of a single-phase grid-connected induction generator with the Steinmetz connection'. IEEE Transactions on Energy Conversion, **18**, 2003; 321–329. © (2003) IEEE*

	$n = 1511$ r/min		$n = 1536$ r/min		$n = 1547$ r/min	
Variables	SYM	FEM	SYM	FEM	SYM	FEM
V_B	0.011	0.077	0.013	0.038	0.027	0.005
V_C	0.02	0.007	0.016	0.01	0.043	0.039
I_A	0.29	0.62	0.23	0.32	0.45	0.11
I_B	0.11	0.11	0.06	0.19	0.02	0.15
I_C	0.20	0.15	0.14	0.01	0.21	0.032
I_L	0.54	0.51	0.25	0.09	0.38	0.15
P_{out}	0.74	0.56	0.26	0.03	0.34	0.14

corrected to that corresponding to the average operating temperature under experimental conditions, but there might be hot spots where the temperatures deviate from the average value, hence causing the generator performance to change. For a more accurate analysis, an appropriate thermal model has to be incorporated with the FEM field analysis model [4].

4. Use of a three-dimensional (3-D) FEM will yield a better modelling of the skewed rotor cage and hence will improve the computational accuracy, particularly when stray losses and efficiency are to be considered [13].

5. In the present rotor circuit model, each rotor cage bar and end ring are partitioned into six layers in order to account for the rotor current distribution. The accuracy will be improved if more layers are used, but more circuit variables will be introduced, thereby increasing the computer memory required.

4.7 Summary

A new approach for analysing the performance of a grid-connected single-phase IG with the Steinmetz connection has been presented in this chapter. A coupled circuit and field approach based on the two-dimensional FEM is adopted in order to account for the asymmetrical stator winding connection as well as the complex magnetic field in the machine and the distribution of current in the rotor winding. Detailed derivation of the stator and rotor circuit equations suitable for coupling with the FEM field equations is given. Experimental results obtained on a small induction machine have also been presented to check the accuracy of both the FEM and symmetrical components analysis. To improve the computational accuracy, a larger number of nodes and elements have to be used for the FEM mesh, in particular for the air gap region. Since the proposed method is based on rigorous machine modelling and is very general, it can be applied to other asymmetrical IG configurations, with appropriate modification in accordance with specific circuit constraints. With the increasing availability of powerful computers at modest cost, the proposed method should be a better alternative for the analysis of single-phase IGs.

References

[1] V. Ostovic, *Dynamics of Saturated Electric Machines*, Springer-Verlag, New York, 1989.

[2] T.F. Chan, 'Performance analysis of a three-phase induction generator connected to a single-phase power system,' *IEEE Transactions on Energy Conversion*, Vol. 13, No. 3, pp. 205–211, September 1998.

[3] N.A. Demerdash and P. Baldassari, 'A combined finite element state space modelling environment for induction motors in the ABC frame of reference: the no load condition,' *IEEE Transactions on Energy Conversion*, Vol. 7, No. 4, pp. 698–709, December 1992.

[4] C.C. Chan, L.-T. Yan, P.-Z. Chen, Z.-Z. Wang and K.T. Chau, 'Analysis of electromagnetic and thermal fields for induction motors during starting,' *IEEE Transactions on Energy Conversion*, Vol. 9, No. 1, pp. 53–60, March 1994.

[5] J. Brauer, H. Sadeghi and R. Oesterlei, 'Polyphase induction motor performance computed directly by finite elements,' *IEEE Transactions on Energy Conversion*, Vol. 14, No. 3, pp. 583–588, September 1999.

[6] T.H. Pham, P.F. Wendling, S.J. Salon and H. Acikgoz, 'Transient finite element analysis of an induction motor with external circuit connections and electromechanical coupling,' *IEEE Transactions on Energy Conversion*, Vol. 14, No. 4, pp. 1407–1412, December 1999.

[7] N.A. Demerdash, J.F. Bangura and A.A. Arkadan, 'A time-stepping coupled finite element–state space model for induction motor drives–Part 1: model formulation and machine parameter computation,' *IEEE Transactions on Energy Conversion*, Vol. 14, No. 4, pp. 1465–1471, December 1999.

[8] N.A. Demerdash and T.W. Nehl, 'Electric machinery parameters and torques by current and energy perturbations from field computations – Part I: theory and formulation,' *IEEE Transactions on Energy Conversion*, Vol. 14, No. 4, pp. 1507–1513, December 1999.

[9] A. Domijan, Jr and Y. Yin, 'Single phase induction machine simulation using the electromagnetic transients program: theory and test cases,' *IEEE Transactions on Energy Conversion*, Vol. 9, No. 3, pp. 535–542, September 1994.

[10] R. Hung and H.W. Dommel, 'Synchronous machine models for simulation of induction motor transients,' *IEEE Transactions on Energy Conversion*, Vol. 11, No. 2, pp. 833–838, May 1996.

[11] S.J. Salon, *Finite Element Analysis of Electric Machines*, Kluwer Academic, Boston, 1995.

[12] Pi-Zhang Chen, Lie-Tong Yan and Ruo-Ping Yao, *Theory and Computation of Electromagnetic Fields in Electrical Machines*, Science Press, China, 1986.

[13] S.L. Ho, W.N. Fu and H.C. Wong, 'Estimation of stray losses of skewed induction motors using coupled 2-D and 3-D time stepping finite element methods,' *IEEE Transactions on Magnetics*, Vol. 34, No. 5, pp. 3102–3105, September 1998.

5

SEIGs for Autonomous Power Systems

5.1 Introduction

This chapter discusses the various circuit configurations for self-excited induction generators (SEIGs) used in autonomous (also known as stand-alone or isolated) power systems. The Steinmetz connection as shown in Figure 5.1 is first considered in order to set forth the basic method of analysis that can be readily extended to more complicated circuit configurations, in particular the general case of asymmetrically connected excitation capacitances and load impedances.

Since the SEIG supplies isolated loads, the frequency of the output voltage is variable even when the rotor speed is maintained constant. To simplify the analysis, all the circuit parameters have been referred to the base (rated) frequency f_{base} by introducing the per-unit frequency a and the per-unit speed b, as explained in Appendix B. Thus, the per-unit slip of the SEIG is $(a - b)/a$ and each voltage shown in Figure 5.1 has to be multiplied by a in order to give the actual value. The *motor* convention will be adopted for the direction of currents in the equivalent circuits.

To illustrate the feasibility of the proposed analysis method and solution technique, a number of case studies were carried out on a 2.2 kW, delta-connected induction machine whose technical data are given in Appendix D.1.

5.2 Three-Phase SEIG with the Steinmetz Connection

5.2.1 Circuit Connection and Analysis

Figure 5.1 shows the Steinmetz connection for a delta-connected three-phase IG self-excited with a single capacitance and supplying a single-phase load, the

Distributed Generation: Induction and Permanent Magnet Generators   L. L. Lai and T. F. Chan
© 2007 John Wiley & Sons, Ltd

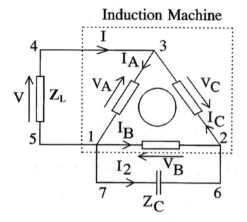

Figure 5.1   *Steinmetz connection for a three-phase SEIG supplying a single-phase load*

capacitance being connected across the lagging phase [1]. The single-phase load is connected across A-phase (the reference phase), while the excitation capacitance C is connected across B-phase (the lagging phase). Besides providing the reactive power for initiating and sustaining self-excitation, C (or the corresponding complex admittance Z_C) also acts as a phase balancer by injecting a line current I_2 into the 'free' terminal of the stator winding. Voltage build-up in the single-phase SEIG is easily initiated by first switching off the load and using a sufficiently large value of C. The stator phase voltages and currents are unbalanced due to the asymmetrical winding configuration, but the phase balance is improved as the load is increased. In case the generator voltage fails to build up due to a previous de-excitation operation, the residual flux should first be re-established by circulating DC momentarily through the stator winding. The self-excitation performance is also improved by operating the generator initially at a higher speed or by using a larger value of excitation capacitance.

With reference to Figure 5.1, the 'inspection equations' may be written as follows:

$$V = V_A \tag{5.1}$$

$$V_A + V_B + V_C = 0 \tag{5.2}$$

$$I_2 = V_B/Z_C = I_C - I_B \tag{5.3}$$

$$I = I_A - I_C = -V/Z_L \tag{5.4}$$

where

$$Z_L = \frac{R_L}{a} + jX_L. \tag{5.5}$$

The inspection equations may be solved using the method of symmetrical components to give the positive- and negative-sequence voltages:

$$V_p = \sqrt{3}V . \frac{Z_p \left(Z_C + \frac{e^{j\pi/6}}{\sqrt{3}} Z_n \right)}{Z_p Z_n + Z_p Z_C + Z_n Z_C} \tag{5.6}$$

$$V_n = \sqrt{3}V . \frac{Z_n \left(Z_C + \frac{e^{-j\pi/6}}{\sqrt{3}} Z_p \right)}{Z_p Z_n + Z_p Z_C + Z_n Z_C} \tag{5.7}$$

where Z_p and Z_n are respectively the positive- and negative-sequence impedances of the generator and $Z_C = 1/(j2\pi f_{base} C.a^2)$ is the complex impedance of the excitation capacitance at the base frequency.

From (5.6) and (5.7) and referring to Figure 5.1, the input impedance of the IG across terminals 1 and 3 is given by

$$Z_{in} = R_{in} + jX_{in} = \frac{Z_p Z_n + Z_p Z_C + Z_n Z_C}{3Z_C + Z_p + Z_n}. \tag{5.8}$$

From (5.8), the SEIG system of Figure 5.1 may be reduced to the simple circuit shown in Figure 5.2.

Applying Kirchhoff's voltage law to the latter circuit,

$$I(Z_L + Z_{in}) = 0. \tag{5.9}$$

For successful voltage build-up, $I \neq 0$, hence

$$Z_{in} + Z_L = 0 \tag{5.10}$$

i.e. the impedances in loop 1345 in Figure 5.2 must sum to zero.

The complex equation (5.10) must be solved in order to give the excitation frequency a and the magnetizing reactance X_m. After a and X_m have been determined, the positive-sequence air gap voltage is found from the magnetization curve. The generator performance can then be computed using (5.1) to (5.7) together with the symmetrical component equations.

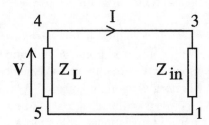

Figure 5.2   *Simplified circuit of three-phase SEIG with Steinmetz connection*

5.2.2   Solution Technique

For a given per-unit speed b and a given excitation capacitance, Z_{in} is a highly nonlinear function of a and X_m, implying that (5.10) is a complex equation in these two variables and hence is extremely difficult to solve using conventional techniques such as the Newton–Raphson method [2] or the polynomial method [3]. To avoid the lengthy mathematical manipulations involved, the solution of (5.10) is formulated as the following optimization problem:

$$\text{Minimize}\quad Z(a, X_m) = \sqrt{\left(R_{in} + \frac{R_L}{a}\right)^2 + (X_{in} + X_L)^2} \qquad (5.11)$$

subject to the constraints:

$$0 < a < b$$

and

$$0 < X_m < X_{mu}$$

where X_{mu} is the unsaturated value of the magnetizing reactance. This approach is based on the fact that, for given values of a and X_m, the input impedance Z_{in} can be readily computed.

It is easy to show that (5.10) is satisfied when the scalar impedance function $Z(a, X_m)$ given by (5.11) assumes a minimum value of zero. For function minimization, a classical search algorithm such as the Hooke and Jeeves method [4] or a commercial optimization package [5] may be employed.

In this and subsequent sections, the pattern search method of Hooke and Jeeves will be used for function minimization for the following reasons:

1. It is a well-proven and robust method that suits a wide range of mathematical functions.
2. This method does not involve derivatives of functions and as such is suitable for problems in which the derivatives of functions do not exist or are difficult to evaluate.
3. The method is relatively simple to program.

This method employs two search strategies, namely *exploratory moves* and *pattern moves*, in order to arrive at the optimum point. A function evaluation is required each time an exploratory move or pattern move is to be made. Details of the method are given in Appendix B. For normal operation of an SEIG, a is slightly less than the per-unit speed b while X_m is less than the unsaturated value X_{mu}, hence b and X_{mu} could in general be chosen as initial estimates for a and X_m for starting the search procedure. After a and X_m have been determined, the steady-state performance can be calculated using the circuit equations and the magnetization curve of the IG.

5.2.3   Capacitance Requirement

This section addresses the capacitor sizing problem of the three-phase SEIG with the Steinmetz connection. A solution method is developed for computing the generator performance and capacitance requirement, taking into account the effect of load impedance, power factor and rotor speed. The solution technique is further extended to the computation of capacitance required to maintain the terminal voltage at a preset value when the generator is on load. The theoretical results are validated by experiments on machine IG1 whose data are given in Appendix D.1.

For a given rotor speed, load impedance and power factor, the value of excitation capacitance must lie within certain limits for an IG to self-excite and to secure a stable operating point. The limiting conditions are obtained when the magnetizing reactance X_m is equal to the unsaturated value X_{mu} [6, 7]. For the Steinmetz connection shown in Figure 5.1, a simple method to determine the capacitance for initiating self-excitation is to consider the impedances in mesh 1267. Using the symmetrical components analysis presented in Section 5.2.1, the input impedance of the generator Z'_{in} when viewed across terminals 1 and 2 is determined as follows:

$$Z'_{in} = R'_{in} + jX'_{in} = \frac{Z_p Z_n + Z_p Z_L + Z_n Z_L}{3Z_L + Z_p + Z_n} \qquad (5.12)$$

where Z_L is given by (5.5). For a given speed and load, both R'_{in} and X'_{in} are functions of the per-unit frequency a only.

Successful voltage build-up requires the sum of impedances in mesh 1267 to be equal to zero, i.e.

$$Z'_{in} + Z_C = 0. \qquad (5.13)$$

The complex equation (5.13) can be split into two algebraic equations by equating the sum of real and imaginary terms respectively to zero:

$$R'_{in}(a) = 0 \qquad (5.14)$$

$$X'_{in}(a) - \frac{X_C}{a^2} = 0 \qquad (5.15)$$

where $X_C = 1/(j2\pi f_{base}C)$.

Equation (5.14) may be solved to yield the per-unit frequency a. The corresponding value of excitation capacitance can then be determined from (5.15).

Equation (5.14) is a high-order polynomial in a. Although the coefficients of the polynomial could be evaluated by systematic algebraic manipulations, the effort required is tremendous. To overcome this difficulty, the secant method [8] is employed in the solution procedure. This method involves the evaluation of function values only and hence is easy to implement. The secant formula as applied to the present problem is

$$a_{n+1} = a_n - \frac{(a_n - a_{n-1})R'_{in}(a_n)}{R'_{in}(a_n) - R'_{in}(a_{n-1})} \qquad (5.16)$$

where a_{n-1}, a_n and a_{n+1} are the values of the per-unit frequency at the end of the $(n-1)$th, nth and $(n+1)$th iteration respectively. The convergence criterion is that the value of per-unit frequency in successive iterations is less than a specified value, say 1.0e-6.

Two initial estimates of a are required for starting the secant method. To ascertain the appropriate initial values, the function R'_{in} is plotted in Figure 5.3 for the experimental machine. It is assumed that the generator is driven at rated speed and is supplying a single-phase load at a power factor (p.f.) of unity. For a given load impedance, there are in general two roots to (5.14). Both roots are less than the per-unit speed b. The lower root a_{min} lies between $0.75b$ and $0.83b$, while the upper root a_{max} lies between $0.83b$ and b. To compute a_{min}, the initial estimates of a can conveniently be chosen as $0.65b$ and $0.7b$. To compute a_{max}, the initial estimates can be chosen as b and $0.99b$.

When the load impedance is very small (e.g. $Z_L = 0.01$ p.u.), the input resistance R'_{in} is always positive and no solution to (5.14) exists. There is thus a critical value of load impedance below which self-excitation is impossible, irrespective of the value of the excitation capacitance.

Figure 5.4 shows the variation of X'_{in} with a. A comparison between Figure 5.3 and Figure 5.4 shows that, for a given load impedance and speed, X'_{in} is in general a monotonously increasing function over the interval $[a_{min}, a_{max}]$. Thus, a_{min} corresponds to C_{max} while a_{max} corresponds to C_{min}. C_{min} is the minimum capacitance required for self-excitation, while C_{max} is the capacitance above which self-excitation is not possible.

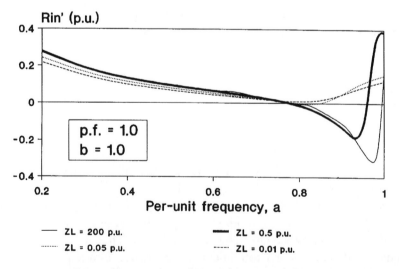

Figure 5.3   *Variation of R'_{in} with per-unit frequency*

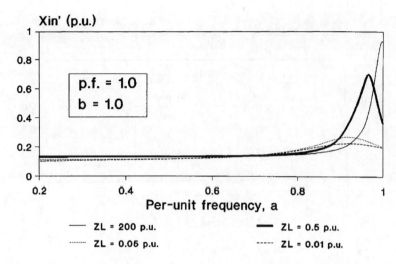

Figure 5.4 Variation of X'_{in} with per-unit frequency. Reproduced by permission of T. F. Chan and L. L. Lai, 'Capacitance requirements of a three phase induction generator self-excited with a single capacitance and supplying a single-phase load', Transactions on Energy Conversion, **17**, 2002; 90–94. © (2002) IEEE

5.2.4   Computed and Experimental Results

A computer program was developed for computing the capacitance requirement of the experimental three-phase SEIG with Steinmetz connection. Typical computed results are listed in Table 5.1.

The secant method is very efficient: convergence can in general be obtained in three to seven iterations over a wide range of load impedance and the function minima are very close to zero. It should be pointed out, however, that the number of iterations and convergence depend on the circuit conditions. For low speeds or small load impedances, the initial values may need to be adjusted in order to speed up the convergence process.

Figure 5.5 shows the computed variation of per-unit frequency and excitation capacitance with load impedance at rated speed. As explained in Section 5.2.3, there are two frequencies, a_{max} and a_{min}, for a given load impedance Z_L and speed, provided that Z_L is greater than some critical value $Z_{L,crit}$. When the load

Table 5.1   Computed results using the secant method ($b = 1.0$; p.f. $= 1.0$; $a_0 = 1.0$; $a_1 = 0.99$)

Z_L (p.u.)	a_{max}	C_{min} (μF)	Number of iterations	Function minimum
200	0.994 886	85.27	4	−3.678e-6
20	0.993 951	85.57	4	−2.225e-6
5	0.990 879	86.81	3	−1.196e-6
2	0.984 949	90.35	5	1.184e-8
1	0.975 696	98.98	7	−3.631e-6

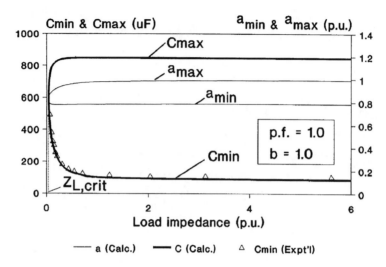

Figure 5.5 *Variation of per-unit frequency and excitation capacitance with load impedance. Reproduced by permission of T. F. Chan and L. L. Lai, 'Capacitance requirements of a three phase induction generator self-excited with a single capacitance and supplying a single-phase load', Transactions on Energy Conversion, 17, 2002; 90–94. © (2002) IEEE*

impedance is equal to $Z_{L,crit}$, there is only one solution to (5.9) and $a_{max} = a_{min}$. For the experimental machine, $Z_{L,crit}$ is equal to 0.0415 p.u. This implies that an SEIG with Steinmetz connection is capable of self-excitation even with a very small load impedance. The variation of a_{max} and a_{min}, and hence C_{min} and C_{max}, is small when the load impedance exceeds 1.0 p.u. But for smaller load impedances, C_{min} increases while C_{max} decreases rapidly with decrease in Z_L.

The experimental results for C_{min} are also given in Figure 5.5. In all cases, the experimental values are slightly higher than the theoretical values but correlation is good.

Figure 5.6 and Figure 5.7 show the computed variation of pre-unit frequency and excitation capacitance with the per-unit speed for given load impedances. The following observations can be made:

1. For a given load impedance, there exists a critical speed b_{crit} below which self-excitation is impossible. The value of b_{crit} increases with decrease in load impedance. When Z_L is varied from 200 p.u. to 0.1 p.u., b_{crit} increases from 0.2 p.u. (point b_{c1} in Figure 5.5) to 0.25 p.u. (point b_{c2}).
2. C_{min} in general increases with decrease in the per-unit speed b.
3. At higher speeds, C_{max} also increases with decrease in b, but below a certain speed (e.g. point P on the computed curve for $Z_L = 200$ p.u. in Figure 5.7), C_{max} decreases with decrease in b.
4. The self-excitation region bounded by C_{min} and C_{max} decreases with decrease in load impedance.
5. The effect of load impedance on C_{min} is much more pronounced than that on C_{max}.

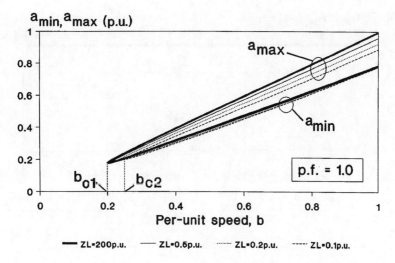

Figure 5.6 *Variation of a_{min} and a_{max} with speed for given load impedances. Reproduced by permission of T. F. Chan and L. L. Lai, 'Capacitance requirements of a three phase induction generator self-excited with a single capacitance and supplying a single-phase load',* Transactions on Energy Conversion, *17, 2002; 90–94. © (2002) IEEE*

The experimental values of C_{min} as a function of per-unit speed are also shown in Figure 5.7. The close agreement between the computed and experimental results in Figure 5.5 and Figure 5.7 confirms the accuracy of the theoretical analysis and solution method.

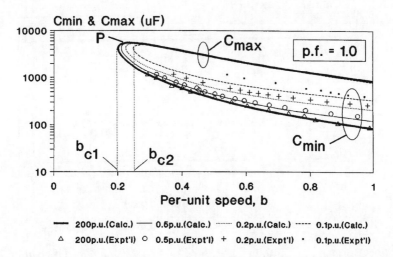

Figure 5.7 *Variation of C_{min} and C_{max} with speed for given load impedances. Reproduced by permission of T. F. Chan and L. L. Lai, 'Capacitance requirements of a three phase induction generator self-excited with a single capacitance and supplying a single-phase load',* Transactions on Energy Conversion, *17, 2002; 90–94. © (2002) IEEE*

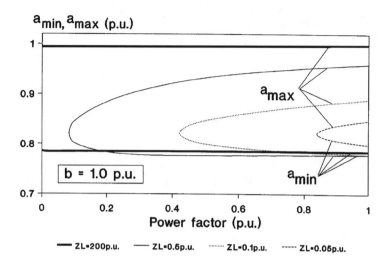

Figure 5.8   *Variation of a_{min} and a_{max} with power factor. Reproduced by permission of T. F. Chan and L. L. Lai, 'Capacitance requirements of a three phase induction generator self-excited with a single capacitance and supplying a single-phase load',* Transactions on Energy Conversion, *17, 2002; 90–94. © (2002) IEEE*

Figure 5.8 and Figure 5.9 show the computed variation of per-unit frequency and excitation capacitance with load power factor. Critical values of power factor, below which the SEIG fails to self-excite, are found to exist. A small load impedance in general requires a high load power factor for voltage build-up.

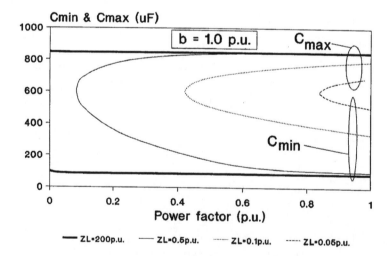

Figure 5.9   *Variation of C_{min} and C_{max} with power factor. Reproduced by permission of T. F. Chan and L. L. Lai, 'Capacitance requirements of a three phase induction generator self-excited with a single capacitance and supplying a single-phase load',* Transactions on Energy Conversion, *17, 2002; 90–94. © (2002) IEEE*

5.2.5   Capacitance Requirement on Load

The capacitances computed in the previous sections are those that mark the onset of self-excitation. When the SEIG is loaded, the load current produces a demagnetizing effect on the air gap field and the terminal voltage drops. To maintain the terminal voltage constant, the excitation capacitance has to be increased as the load current increases. On the basis of the analysis presented in the previous sections, it is possible to compute the capacitance requirement of the SEIG under loaded conditions using an iterative procedure which is summarized as follows:

1. For a specified terminal voltage V, power factor and per-unit speed b, assume the following initial values of per-unit frequency and positive-sequence air gap voltage:

$$a_0 = b; E_1 = V/a_0.$$

2. Obtain the corresponding value of magnetizing reactance using the $X_m - E_1$ curve which can be derived from the magnetization characteristic determined in a synchronous speed test.
3. Using the above values, determine the generator impedance $Z'_{in} = R'_{in} + jX'_{in}$ across terminals 4 and 5 in the circuit shown in Figure 5.1.
4. Determine the values of per-unit frequency a and capacitive reactance X_c that satisfy the self-excitation conditions using the secant method.
5. Compute the positive-sequence component of voltage V_p using (5.6).
6. Compute the new value of air gap voltage E_1 from the positive-sequence equivalent circuit.
7. Repeat steps 1 to 6 until the difference in V_p in successive iterations is sufficiently small.
8. Compute the excitation capacitance and the generator performance using the final values of a and X_C.

Since only a_{max} needs to be computed, initial values of a required to start the secant method can be chosen as $0.99b$ and $0.98b$. In general, the number of iterations increases with decrease in load impedance. For the experimental machine IG1 and with a tolerance of 1.0e-4 as the convergence criterion (step 7), the number of iterations varies from 7 to 15, showing that the above iterative procedure converges quite rapidly.

Figure 5.10 shows the excitation capacitance required to maintain a specified terminal voltage when the SEIG is supplying a unity power factor load at rated speed. There is good agreement between the computed and experimental results when the terminal voltage is 0.8 p.u. For a terminal voltage of 1.0 p.u., however, the discrepancy between computed and experimental results increases with the load current. This could be attributed to the more pronounced saturation and hence more complicated magnetic conditions within the machine.

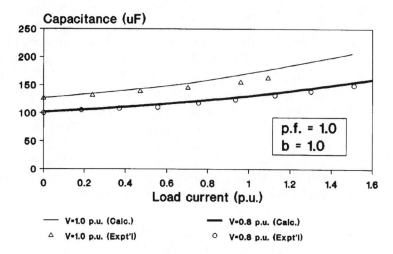

Figure 5.10   *Excitation capacitance required to maintain the terminal voltage at preset values. Reproduced by permission of T. F. Chan and L. L. Lai, 'Capacitance requirements of a three phase induction generator self-excited with a single capacitance and supplying a single-phase load',* Transactions on Energy Conversion, *17, 2002; 90–94. © (2002) IEEE*

Figure 5.11 shows the computed capacitance required to maintain the terminal voltage at 1.0 p.u. at different load power factors. The capacitance required increases rapidly when the power factor is lower than 0.8 lagging.

Figure 5.10 and Figure 5.11 suggest that a controller for varying the excitation capacitance should be incorporated in the SEIG system for stabilization of the load voltage.

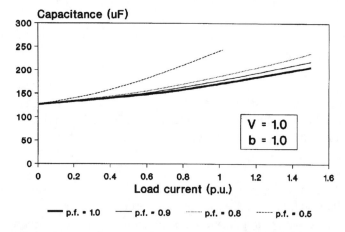

Figure 5.11   *Effect of power factor on the capacitance required to maintain the terminal voltage at rated value. Reproduced by permission of T. F. Chan and L. L. Lai, 'Capacitance requirements of a three phase induction generator self-excited with a single capacitance and supplying a single-phase load',* Transactions on Energy Conversion, *17, 2002; 90–94. © (2002) IEEE*

5.2.6 Summary

This section has presented the analysis of a three-phase SEIG with Steinmetz connection. The method of symmetrical components is used in association with a function optimization method in order to account for the circuit asymmetry and the variation of machine parameters. Based on this approach, a simple method for capacitor sizing has been developed. The method involves the derivation of the generator input impedance and the formulation of two nonlinear equations from a consideration of the self-excitation conditions. By solving one of the equations using the secant method, the per-unit frequency is determined and the excitation capacitance can subsequently be computed using the second equation. The effects of load impedance, speed and power factor on the capacitance requirement have been studied. Critical values of load impedance, speed and power factor, below which the SEIG fails to self-excite, are found to exist. Compared with the single-phasing mode of operation, however, the SEIG with Steinmetz connection can self-excite even with very small load impedance. An iterative procedure is developed for computing the capacitance required to maintain a preset terminal voltage when the SEIG is on load. Where possible, the theoretical analysis has been verified by experiments on a small induction machine.

5.3 SEIG with Asymmetrically Connected Impedances and Excitation Capacitances

When the power rating of a three-phase SEIG becomes smaller, it is increasingly difficult to ensure an even distribution of the loads among the phases, which means that in general the SEIG has to operate with a certain degree of phase imbalance. Unbalanced operation may also result due to manufacturing tolerances of the excitation capacitances, failure of some excitation capacitance modules or the disconnection of loads by consumers.

Autonomous power systems often employ single-phase distribution schemes for reasons of low cost, ease of maintenance and simplicity in protection [9]. When a three-phase SEIG is used to supply single-phase loads, the inherent phase imbalance in the machine will result in poor generator performance, such as overcurrent, overvoltage, poor efficiency, excessive temperature rise and machine vibration. These undesirable effects can be alleviated to a large extent by the use of the Steinmetz connection as discussed in Section 5.2. For isolated operation, however, perfect phase balance cannot be achieved for a pure resistive load [1] or a series *R–L* load [10].

The objectives of this section are to develop a general method for analysing the steady-state performance of a three-phase SEIG under various unbalanced operating conditions and to investigate a novel phase balancing scheme for the SEIG when supplying single-phase loads. Theoretical and experimental results will be compared in order to validate the analysis and the principle of phase balancing.

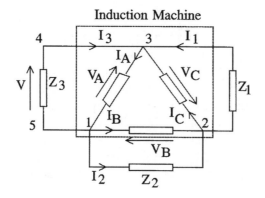

Figure 5.12   *Three-phase SEIG with asymmetrically connected terminal impedances*

5.3.1   Circuit Model

Figure 5.12 shows the circuit connection of a delta-connected IG with asymmetrically connected terminal impedances. At least one of the terminal impedances Z_1, Z_2 and Z_3 must contain a capacitive element in order to furnish the reactive power necessary for initiating self-excitation.

The circuit model shown in Figure 5.12 can be used to study practically all modes of unbalanced operation of the SEIG in which zero-sequence quantities are absent. By assigning appropriate values to the terminal impedances, a specific unbalanced operating condition can be simulated.

A star-connected SEIG can also be analysed by first transforming the generator to an equivalent delta-connected machine whose per-phase impedance is three times the actual star-connected value. In the case of star-connected load impedances and excitation capacitances with isolated neutral points, star–delta transformation can likewise be applied to yield the equivalent delta-connected impedance values. After these transformation procedures, the circuit will be reduced to the generic form as shown in Figure 5.12.

For loads to be supplied by a four-wire system, a delta–star-connected transformer can be placed between the generator and the loads so that zero-sequence currents are excluded from the SEIG. With appropriate impedance transformations, the system is again reduced to that shown in Figure 5.12.

5.3.2   Performance Analysis

The method of symmetrical components is employed in order to account for the unbalanced circuit conditions. All the equivalent circuit parameters are assumed to be constant except the magnetizing reactance, which is a function of the positive-sequence air gap voltage. Since there is no active voltage source, the SEIG may be regarded as a passive circuit when viewed across any two stator terminals. For convenience, A-phase is chosen as the reference and the input impedance of the

SEIG across terminals 1 and 3 in Figure 5.12 will be considered. If V is regarded as an impressed voltage, the system is identical to that shown in Figure 3.1 and hence can be analysed in the same manner as in Section 3.2.

From the analysis, the voltage across terminals 1 and 3 in Figure 5.12 is determined as follows:

$$V = \frac{V_p}{\sqrt{3}}(h - h^2) \cdot \frac{Y_1 + Y_2 + Y_p + Y_n}{hY_2 + (h - h^2)Y_n - h^2 Y_1} \tag{5.17}$$

where h is the complex operator $ej2\pi/3$.

The input current I_3 is given by

$$I_3 = \frac{V_p}{\sqrt{3}}(h - h^2)\frac{(Y_1 + Y_2)(Y_p + Y_n) + 3Y_p Y_n + Y_1 Y_2}{hY_2 + (h - h^2)Y_n - h^2 Y_1}. \tag{5.18}$$

From (5.17) and (5.18), the input impedance Z_{in} of the SEIG when viewed across terminals 1 and 3 is given by

$$Z_{in} = \frac{Y_1 + Y_2 + Y_p + Y_n}{(Y_1 + Y_2)(Y_p + Y_n) + 3Y_p Y_n + Y_1 Y_2}. \tag{5.19}$$

Both Y_p and Y_n are functions of the per-unit frequency a and the magnetizing reactance X_m, hence the input impedance of the SEIG may be written as

$$Z_{in} = R_{in}(a, X_m) + jX_{in}(a, X_m). \tag{5.20}$$

From (5.20), the SEIG system of Figure 5.12 may also be reduced to the simple circuit shown in Figure 5.2, with Z_L replaced by Z_3. For successful voltage build-up,

$$Z_3 + Z_{in} = 0. \tag{5.21}$$

The complex equation (5.21) may be solved using the solution technique described in Section 5.2.2. The scalar impedance function to be minimized is

$$Z(a, X_m) = \sqrt{(R_3 + R_{in})^2 + (X_3 + X_{in})^2} \tag{5.22}$$

where R_3 and X_3 are respectively the equivalent series resistance and reactance of the terminal impedance Z_3. After a and X_m have been determined, the positive-sequence air gap voltage is found from the magnetization curve. The generator performance can then be computed using the circuit equations together with the symmetrical component equations.

5.3.3   Computed and Experimental Results

A number of case studies were carried out on the experimental machine IG1 whose data are given in Appendix D.1. Appropriate numerical values are assigned to the terminal impedances in order to simulate a particular unbalanced operating condition. For example, a zero capacitance value implies the absence of an excitation capacitance while a very large resistance value, say 1.0e5 p.u., is equivalent to the

Table 5.2   *Case studies of three-phase SEIG with asymmetrically connected terminal impedances*

	$Z_1 = R_{L1} \mathbin{/\mkern-5mu/} C_1$		$Z_2 = R_{L2} \mathbin{/\mkern-5mu/} C_2$		$Z_3 = R_{L3} \mathbin{/\mkern-5mu/} C_3$		
Case	R_{L1} (p.u.)	C_1 (μF)	R_{L2} (p.u.)	C_2 (μF)	R_{L3} (p.u.)	C_3 (μF)	Remarks
1	2.10	60	1.10	60	5.0	60	Balanced capacitances; unbalanced load resistance
2	2.15	30	2.15	60	2.15	60	Unbalanced capacitance; balanced load resistance
3	2.95	60	3.30	62	2.15	72	Unbalanced capacitance; unbalanced load resistance
4	∞	60	∞	60	0.97	60	Balanced capacitance; single-phase load resistance
5	∞	0	∞	0	1.65	125	Plain single-phasing operation
6	∞	0	∞	125	0.90	0	Steinmetz connection

absence of a connected load resistance. Operation at rated speed is assumed in each case investigated.

To simplify the calculations and for easy comparison, all the parameters are expressed in per-unit values, using the rated phase voltage, rated phase current and rated power per phase of the induction machine as bases. Over a wide range of load and for various unbalanced cases, convergence can be obtained in 350 to 450 function evaluations for the experimental machine. Table 5.2 summarizes the cases studied and shows the typical circuit parameters used. As shown in Table 5.3, very good agreement between the computed and experimental results is obtained for most of the unbalanced cases, hence the accuracy of the analysis and solution procedure is verified. Larger discrepancies exist, however, for case 5. The severe phase imbalance for plain single-phasing operation results in nonlinearity in the magnetic circuit of the SEIG. The principle of superposition, which is the basis of symmetrical components analysis, is therefore less valid.

5.3.4   Modified Steinmetz Connection

Many autonomous electric power systems employ single-phase generation and distribution for reasons of simplicity and reduced cost. An examination of cases 4 and 5 in Tables 5.2 and 5.3, however, reveals that poor generator performance, such as phase imbalance and poor efficiency, will result when single-phase loads are supplied by a three-phase SEIG. These disadvantages are overcome to a large extent by the use of the Steinmetz connection, which has been investigated in considerable detail in Section 5.2. By connecting the excitation capacitance and

Table 5.3   *Performance of SEIG for the circuit configurations listed in Table 5.2 (Normal: experimental values; Bracketed: computed values)*

Case	Phase voltages (p.u.)			Phase currents (p.u.)			VUF^a	P_{out} (p.u.)^a	Efficiency (p.u.)
	V_A	V_B	V_C	I_A	I_B	I_C			
1	1.0091	0.9545	0.9545	0.9574	1.046	0.713	0.0486	1.47	0.7998
	(1.0507)	(0.9808)	(0.9686)	(0.9736)	(1.0648)	(0.7141)	(0.0518)	(1.545)	(0.7747)
2	0.9045	0.8818	0.8727	0.7204	0.7796	0.6167	0.0288	1.0968	0.8174
	(0.9197)	(0.8875)	(0.8733)	(0.7103)	(0.7933)	(0.5926)	(0.0309)	(1.115)	(0.7905)
3	1.0818	1.075	1.0909	1.0278	0.9074	1.011	0.0095	1.299	0.7616
	(1.0902)	(1.0781)	(1.1046)	(1.0154)	(0.9032)	(0.9709)	(0.0141)	(1.3197)	(0.7447)
4	0.9909	1.0364	1.1182	1.0057	0.4852	1.1667	0.0878	1.0093	0.7133
	(1.0057)	(1.0438)	(1.1618)	(1.0192)	(0.4739)	(1.1716)	(0.0896)	(1.0396)	(0.6934)
5	0.824	0.6727	0.8368	0.9444	0.4407	0.4407	0.1321	0.4209	0.6241
	(0.9261)	(0.7337)	(0.9272)	(1.0352)	(0.5176)	(0.5176)	(0.145)	(0.5204)	0.6096
6	0.8291	0.9045	0.8355	0.3296	0.6741	0.7778	0.0573	0.7357	0.7675
	(0.8633)	(0.9547)	(0.867)	(0.3521)	(0.7443)	(0.7812)	(0.0681)	(0.8256)	(0.7521)

^{a}VUF = Voltage Unbalance Factor; P_{out} = total electric power output.

load across different phases, better phase balance and higher efficiency could be obtained and the minimum voltage unbalance factor was about 5% for a pure resistive load.

Subsequent research on the Steinmetz connection for SEIGs reveals that additional circuit elements are required in order to achieve perfect phase balance. Based on this result, a modified Steinmetz connection (MSC) for a three-phase SEIG is proposed in this section. Figure 5.13 shows the circuit connection of the MSC, where all circuit parameters have been referred to the base frequency. The impedance Z_3 across A-phase (the reference phase) consists of the main load resistance R_{L3} and the auxiliary excitation capacitance in parallel. The impedance Z_2 across B-phase (the lagging phase) consists of the main excitation capacitance C_2 and auxiliary load resistance R_{L2} in parallel. Compared with the original Steinmetz

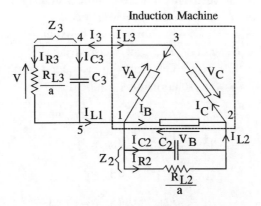

Figure 5.13   *Modified Steinmetz connection for three-phase SEIG*

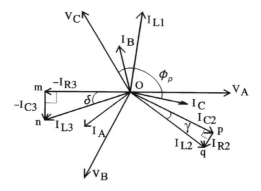

Figure 5.14   *Phasor diagram for SEIG with MSC under perfect phase balance*

connection in Figure 5.1, it is seen that the auxiliary resistance R_{L2} and the auxiliary excitation capacitance C_3 have been introduced. For a practical SEIG system, R_{L2} could be local loads such as lighting, storage heating or battery charging. Alternatively, R_{L2} could be a portion of the remote loads.

The three-phase SEIG with MSC can also be analysed using the general method described in Sections 5.3.2 and 5.3.3. In this case, Y_1 is equal to zero while Y_2 and Y_3 are the resultant admittances connected across B-phase and A-phase respectively.

Conditions for perfect phase balance.   Figure 5.14 shows the phasor diagram of the three-phase SEIG with MSC under balanced conditions, it being assumed that the positive-sequence impedance angle ϕ_p is greater than $2\pi/3$ rad. The line current I_{L2} flowing into terminal 2 consists of the current I_{C2} through the main excitation capacitance C_2 and the current I_{R2} through the auxiliary resistance R_{L2}. Meanwhile, the line current I_{L3} flowing into terminal 3 is contributed by $-I_{R3}$ (where I_{R3} is the main load current) as well as $-I_{C3}$ (where I_{C3} is the current through the auxiliary capacitance C_3). The current components I_{R2} and I_{C3} enable balanced line currents of the SEIG to be synthesized.

A careful study of the relationship between the voltage and current phasors in Figure 5.14 shows that, under perfect phase balance, the angle γ between I_{C2} and I_{L2} is equal to $(\phi_p - 2\pi/3)$ rad while the angle δ between $-I_{R3}$ and I_{L3} is equal to $(5\pi/6 - \phi_p)$ rad. Since each current in the phasor triangles Opq and Omn may be expressed in terms of the phase voltage and the associated admittance, conductance or susceptance, the following relationships can be derived:

$$G_2 = \sqrt{3}|Y_p|\sin(\phi_p - 2\pi/3) \tag{5.23}$$

$$B_2 = \sqrt{3}|Y_p|\cos(\phi_p - 2\pi/3) \tag{5.24}$$

$$G_3 = \sqrt{3}|Y_p|\cos(5\pi/6 - \phi_p) \tag{5.25}$$

$$B_3 = \sqrt{3}|Y_p|\sin(5\pi/6 - \phi_p) \tag{5.26}$$

where $G_2 = a/R_{L2}$, $B_2 = a^2.2\pi f_{base}.C_2$, $G_3 = a/R_{L3}$ and $B_3 = a^2.2\pi f_{base}.C_3$.

Table 5.4   *Circuit conditions to give balanced operation of three-phase SEIG with MSC (Normal: experimental values; Bracketed: computed values)*

				$Z_2 = R_{L2} \,/\!/\, C_2$		$Z_3 = R_{L3} \,/\!/\, C_3$	
V_{ph} (p.u.)	I_{ph} (p.u.)	Y_p (p.u.)	ϕ_p (deg)	R_{L2} (p.u.)	C_2 (μF)	R_{L3} (p.u.)	C_3 (μF)
0.805	0.954	1.214	134.7	1.775 (1.873)	168 (167)	0.5069 (0.493)	49 (46)
0.899	0.878	0.975	130.8	2.88 (3.1)	133 (132)	0.641 (0.614)	45 (44)
0.999	1.006	1.007	125.6	4.91 (5.7)	141 (139)	0.644 (0.615)	58 (57)
0.985	0.768	0.779	120	∞ (∞)	104 (107)	0.872 (0.843)	56 (54)
1.064	1.019	0.957	120	∞ (∞)	134 (134)	0.675 (0.722)	67 (67)

When ϕ_p is greater than $2\pi/3$ rad (which corresponds to a heavy load condition), G_2 is positive and perfect balance can be obtained with all four circuit elements in Figure 5.13 present. When ϕ_p is equal to $2\pi/3$ rad, G_2 vanishes showing that phase balance can be achieved with the auxiliary load resistance removed. Under this condition, $B_2 = \sqrt{3}Y_p$, $B_3 = \sqrt{3}Y_p/2$ and $G_3 = 3Y_p/2$. When ϕ_p is less than $2\pi/3$ rad, however, G_2 is negative and perfect phase balance cannot be obtained with passive circuit elements.

Equation (5.26) shows that B_3 vanishes when $\phi_p = 5\pi/6$ rad, which implies that the auxiliary capacitance C_3 can be dispensed with. When ϕ_p exceeds $5\pi/6$ rad, B_3 becomes negative, implying that perfect balance can be achieved with an auxiliary inductance across A-phase. In practice, however, the full-load power factor angle of an SEIG ranges from $2\pi/3$ to $4\pi/5$ rad, hence it is very unlikely that an inductive element need be used.

To investigate the phase balancing capability of the MSC, experiments were performed on the 2.2 kW induction machine. The rotor speed is maintained at rated value throughout the tests. The values of R_{L2}, C_2, R_{L3} and C_3 were carefully adjusted to give perfect phase balance in the SEIG for specific values of phase current. Typical results are summarized in Table 5.4. The good agreement between the computed and experimental values of the circuit parameters verifies the principle of phase balancing for a three-phase SEIG using the MSC.

Figures 5.15–5.17 show the performance characteristics of the SEIG in which $C_2 = 146\ \mu$F, $C_3 = 47\ \mu$F and $R_{L2} = 94\ \Omega$ (2.3 p.u.). These values of phase converter elements result in perfect phase balance at an experimental load current of 1.52 p.u. and a phase current of 0.93 p.u. When the load is reduced, I_B increases rapidly, while I_C decreases. On the other hand, I_A remains substantially constant for load currents down to 0.8 p.u. Provided that the load does not vary too widely from that corresponding to perfect phase balance, satisfactory performance of the SEIG can still be obtained. Under no-load conditions, however, there will be severe over-current and overvoltage in A-phase and B-phase, hence the excitation capacitances need to be reduced. Figure 5.17 shows that, under perfect phase balance, the SEIG delivers a power of 1.31 p.u. and 0.323 p.u. (experimental values) to the main load and auxiliary load respectively. In other words, about 80 % of the electric power

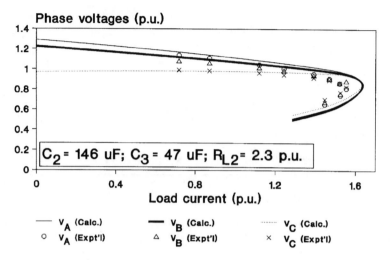

Figure 5.15   *Phase voltages of SEIG with MSC*

output is delivered to the main load. The total load is 1.63 p.u. (1940 W), which is 88 % of the rated power of the induction machine. An experimental efficiency of 80 % is obtainable at and close to the load corresponding to perfect phase balance. Very good agreement between the computed and experimental characteristics has been obtained.

Selection of phase converter elements for a given load.   A practical design problem is, for a given speed and main load resistance R_{L3}, to determine the values of the phase converter elements in order to give perfect phase balance in the three-phase

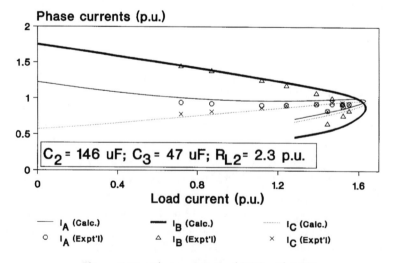

Figure 5.16   *Phase currents of SEIG with MSC*

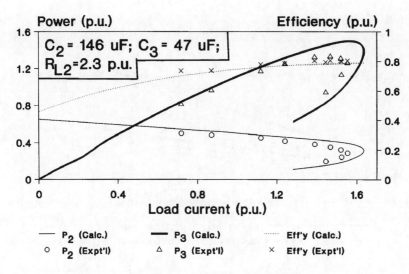

Figure 5.17 *Output power and efficiency of SEIG with MSC (P_2: output power to auxiliary load R_{L2}; P_3: output power to main load R_{L3})*

SEIG. From (5.23)–(5.26), it is observed that G_2, B_2, G_3 and B_3 are all functions of the variables Y_p and ϕ_p which depend on the terminal impedances. An iterative procedure is therefore required to determine the values of the phase converter elements that give perfect phase balance. For convenience, ϕ_p can be specified first while Y_p is to be determined during the iterations. The iterative procedure may be summarized as follows:

1. Input the per-unit speed b and specified value of ϕ_p.
2. Assume an initial value of the per-unit frequency a.
3. For a given value of main load resistance R_{L3}, compute $|Y_p|$ from (5.25) using the current value of a.
4. Compute B_2, G_2 and B_3 using (5.23), (5.24) and (5.26).
5. Compute Y_2 and Y_3 (hence R_3 and X_3) in Figure 5.13, using the values of circuit elements obtained in steps 3 and 4.
6. Determine a and X_m using the Hooke and Jeeves method outlined in Appendix B.
7. Repeat steps 3 to 6 until the values of a in successive iterations are less than a specified value.
8. Compute the values of phase converter elements and performance of the SEIG using the final values of a and X_m.

The above procedure has been tested with reference to the experimental machine IG1. Convergence can usually be obtained in three to five iterations. Figure 5.18 and Figure 5.19 show the computed values of C_2, C_3 and R_{L2} to give perfect phase balance in the SEIG for given values of main load conductance G_{L3}. From Figure 5.20,

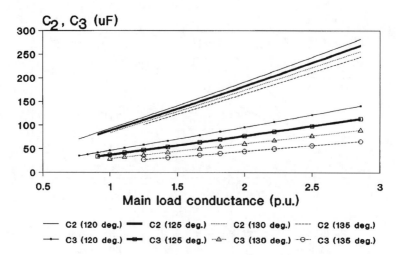

Figure 5.18   *Main and auxiliary excitation capacitances for perfect phase balance in SEIG with MSC*

it is observed that both the phase voltage and phase current increase with decrease in ϕ_p until the limiting value of 120°e (electrical degrees) is reached. The SEIG is thus more likely to experience overvoltage and overcurrent at small values of ϕ_p.

Figure 5.21 shows the total output power and efficiency of the SEIG with MSC under balanced conditions. At $\phi_p = 120°$e, rated current occurs when $G_{L3} = 1.4$ p.u. and the total output power is 1.6 p.u. At $\phi_p = 135°$e, rated current occurs when $G_{L3} = 1.88$ p.u. and the total output power is 1.86 p.u. From voltage, current, output power and efficiency considerations, it is desirable to operate the SEIG at higher values of ϕ_p.

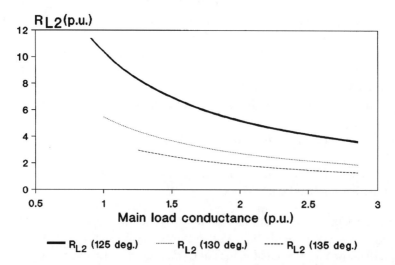

Figure 5.19   *Auxiliary load resistance for perfect phase balance in SEIG with MSC*

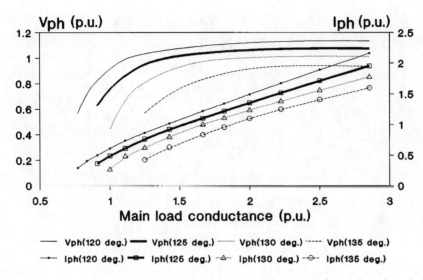

Figure 5.20   *Phase voltage and phase current of SEIG with MSC under perfect phase balance*

5.3.5   Simplified Steinmetz Connection

In circumstances where it is not practicable to provide auxiliary loads, or when auxiliary loads need not be supplied, the simplified Steinmetz connection (SSC) shown in Figure 5.22 may be employed. In this case, all the electric power output of the SEIG is delivered to the single-phase load R_{L3}. The phasor diagram for the MSC (Figure 5.13) and the corresponding equations (5.23)–(5.26) may be used to identify the conditions for perfect phase balance for the SSC. Since the auxiliary

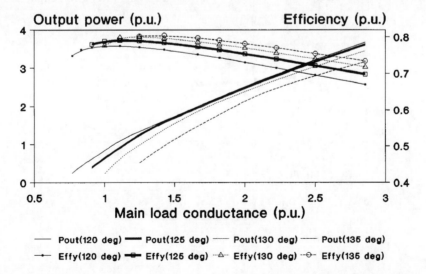

Figure 5.21   *Output power and efficiency of SEIG with MSC under perfect phase balance*

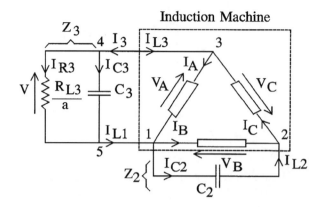

Figure 5.22   *Simplified Steinmetz connection for three-phase SEIG*

load resistance R_{L2} is absent, the value of G_2 in (5.23) is forced to assume a zero value. Accordingly the positive-sequence impedance angle ϕ_p of the SEIG must be equal to $2\pi/3$ rad for (5.23) to be satisfied. From (5.24), (5.25) and (5.26), the values of the load conductance and phase converter susceptances that result in balanced operation of the SEIG are: $B_2 = \sqrt{3}Y_p$, $G_3 = 3Y_p/2$ and $B_3 = \sqrt{3}Y_p/2$.

The auxiliary excitation capacitance C_3 is thus one-half of the main excitation capacitance C_2. By selecting proper values of C_2 and C_3, perfect phase balance can be achieved for a specific value of stator current.

Analysis of the SEIG with SSC is similar to that for the SEIG with MSC, except that the admittance Y_2 is now equal to $(0 + jB_2)$.

Figures 5.23–5.25 show the computed and experimental performance of the SEIG with SSC at rated speed. With the main and auxiliary excitation capacitances

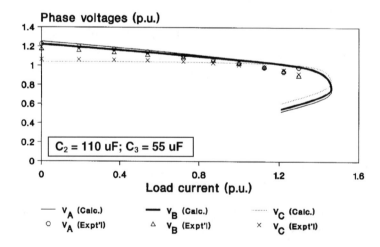

Figure 5.23   *Phase voltages of three-phase SEIG with SSC*

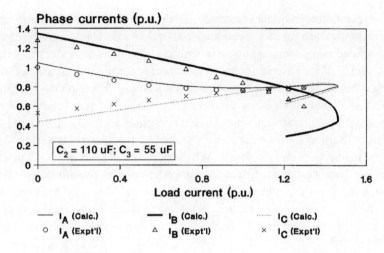

Figure 5.24 *Phase currents of three-phase SEIG with SSC*

fixed at $110\,\mu F$ and $55\,\mu F$ respectively, the SEIG is balanced at a load current (experimental value) of 1.13 p.u, which corresponds to a phase voltage of 0.985 p.u. and a phase current of 0.77 p.u. Under this condition, a power of 1.11 p.u. (1320 W) is delivered to the load and the efficiency of the SEIG is 79.6 %. Again very good agreement between the computed and experimental results is observed.

5.3.6 Summary

A general analysis for a three-phase SEIG with asymmetrically connected load and excitation capacitances has been presented. The equivalent circuit variables,

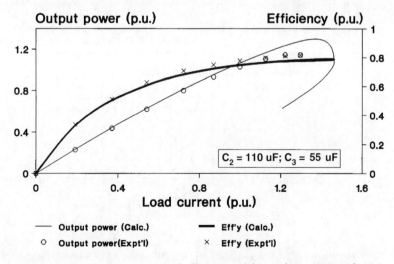

Figure 5.25 *Output power and efficiency of three-phase SEIG with SSC*

namely the excitation frequency and magnetizing reactance, are determined by the function minimization method presented in Section 5.2. The close agreement between computed and experimental results obtained verifies the theory and solution method for an SEIG with asymmetrically connected terminal impedances. A phase balancing scheme for a three-phase SEIG supplying a single-phase load, namely the MSC, has also been investigated. From the voltage/current relationship in the phasor diagram, the conditions for balanced operation are deduced and an iterative method to determine the corresponding values of the phase converter elements for a given load has been developed. When there is no auxiliary load resistance, perfect phase balance can still be achieved provided that the auxiliary excitation capacitance is half of the main excitation capacitance. Since the MSC or the SSC scheme involves only passive circuit elements, it is an economical and effective method for achieving perfect phase balance in a three-phase SEIG that supplies single-phase loads.

5.4 Self-regulated SEIG for Single-Phase Loads

5.4.1 Circuit Connection and Analysis

In this section, series capacitance compensation will be applied to a three-phase SEIG with the Steinmetz connection to give a single-phase, self-regulated, self-excited induction generator (SRSEIG) with reduced voltage regulation, better phase balance and increased power output. Further, it will be shown that a condition of perfect phase balance in the three-phase machine can be achieved over a wide range of load with this new excitation scheme.

Figure 5.26 shows the circuit connection of the single-phase SRSEIG for a delta-connected induction machine. The shunt excitation capacitance C_{sh} is selected to

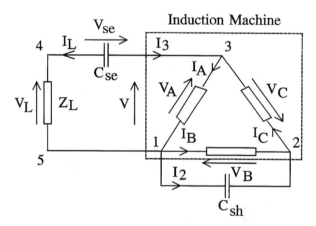

Figure 5.26 *Circuit connection of single-phase SRSEIG using a three-phase delta-connected induction machine*

give the desired no-load voltage, while the series compensation capacitance C_{se} provides additional reactive power when the load current increases, resulting in a reduced voltage drop.

A general analysis of the single-phase SRSEIG can be carried out using the method presented in Section 5.2. For the SRSEIG system, the input impedance Z_{in} as viewed across terminals 1 and 3 is given by

$$Z_{in} = \frac{Z_p Z_n + Z_p Z_{sh} + Z_n Z_{sh}}{3Z_{sh} + Z_p + Z_n} \tag{5.27}$$

where

$$Z_{sh} = \frac{1}{j2\pi f_{base}.C_{sh}.a^2} = -jX_{sh}. \tag{5.28}$$

The complex impedance of the series capacitance C_{se} is

$$Z_{se} = \frac{1}{j2\pi f_{base}.C_{se}.a^2} = -jX_{se}. \tag{5.29}$$

The following scalar impedance function should thus be minimized for solution of the SRSEIG circuit:

$$Z(a, X_m) = \sqrt{\left(R_{in} + \frac{R_L}{a}\right)^2 + (X_{in} + X_L - X_{se})^2}. \tag{5.30}$$

To simplify the calculations and for easy comparison, all the machine parameters are expressed in per-unit values using the rated phase voltage and rated phase current as bases. Table 5.5 shows the computed results for machine IG1 (technical details of which are given in Appendix D.1), with an excitation capacitance of 125 μF and a series compensation capacitance of 350 μF. The per-unit speed and the load power factor are both equal to unity. It is observed that, over a wide range of load impedance, the number of function evaluations N required for a solution varies from 330 to 550. The results here indicate that the Hooke and Jeeves method

Table 5.5 Computed results for single-phase SRSEIG using the Hooke and Jeeves method ($C_{sh} = 125\,\mu F$; $C_{se} = 350\,\mu F$; $b = 1.0$, p.f. = 1.0; $a_0 = 0.97b$; $X_{m0} = X_{mu} = 2.48\,p.u.$)

R_L (p.u.)	a	X_m (p.u.)	Number of function evaluations, N	$Z(a, X_m)$ (p.u.)
50	0.9916	1.6226	381	9.06e-4
10	0.9900	1.6329	332	1.32e-4
5	0.9880	1.6443	390	6.24e-5
2	0.9823	1.6686	396	1.32e-5
1	0.9737	1.6771	356	6.43e-7
0.5	0.9593	1.5982	417	8.53e-7
0.3	0.9455	1.4056	509	2.92e-7
0.1	0.9240	0.8335	386	4.0e-7

is computationally efficient and is suitable for solving the single-phase SRSEIG performance analysis problem. The accuracy is very good as observed from the function minima obtained in Table 5.5.

5.4.2   Effect of Series Compensation Capacitance

It is of interest to investigate the effect of C_{sh} and C_{se} on the performance of the single-phase SRSEIG. In practice, C_{sh} is chosen so as to secure self-excitation and a stable operating point without causing excessive phase voltages and currents at no load. Depending on the voltage regulation and phase balance requirements, different values of C_{se} can be chosen. To facilitate the subsequent discussion, a parameter called the compensation factor K is defined as follows:

$$K = \frac{X_{se}}{X_{sh}} = \frac{C_{sh}}{C_{se}} \tag{5.31}$$

where X_{sh} and X_{se} have been defined in (5.28) and (5.29), respectively.

For the purpose of comparison, it is assumed in this section that the single-phase SRSEIG is operating at rated speed and is supplying a unity-power-factor load.

Voltage regulation. When choosing the value of the series compensation capacitance C_{se}, one should consider the voltage drop across C_{se} as well as the amount of compensating reactive power available. A large value of C_{se} results in a smaller voltage drop, but the reactive power ($I_L^2 X_{se}$) generated is also small. On the other hand, a small value of C_{se} results in a larger voltage drop but provides more reactive power for voltage compensation.

Figure 5.27 shows the computed variation of magnetizing reactance X_m of the experimental SRSEIG with the load admittance Y_L for different values of K when a shunt excitation capacitance of 125 μF is used. Without series compensation

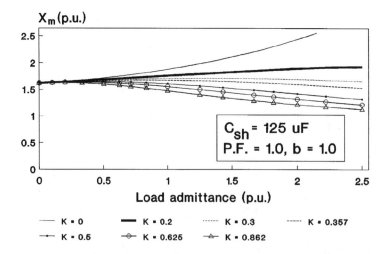

Figure 5.27   *Variation of magnetizing reactance with load admittance*

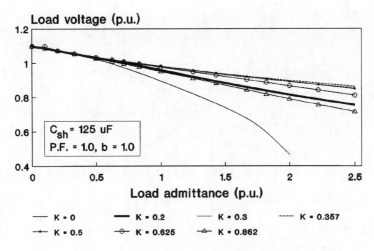

Figure 5.28   *Variation of load voltage with load admittance*

$(K = 0)$, X_m increases rapidly with load conductance and reaches the unsaturated value of 2.48 p.u. when Y_L is equal to 2.05 p.u. At larger values of K, the range of Y_L over which X_m is less than the unsaturated value is extended. This implies that the SRSEIG is capable of maintaining a high terminal voltage over a wider range of load. At $K = 0.357$, X_m remains substantially constant over the practical range of Y_L, i.e. the saturation level of the machine is only slightly affected by the load.

Figure 5.28 shows the computed variation of load voltage with Y_L under the above operating conditions. The best voltage regulation characteristics are obtained when K is between 0.3 and 0.5. When K is less than 0.25 or greater than 0.8, the load voltage decreases significantly.

Phase balancing capability.   One novel feature of the single-phase SRSEIG is that it can operate with balanced phase voltages and currents in the three-phase machine. This occurs when the negative-sequence voltage V_n vanishes. From (5.7) and with Z_C replaced by Z_{sh}, the following condition is deduced:

$$Z_{sh} + \frac{e^{-j\pi/6}}{\sqrt{3}}.Z_p = 0. \tag{5.32}$$

Since the shunt excitation capacitance may be considered to be lossless, from (5.32) one obtains

$$|Z_p| = \sqrt{3}|Z_{sh}| = \sqrt{3}X_{sh} \tag{5.33}$$

and

$$\phi_p = \frac{2\pi}{3} \tag{5.34}$$

where ϕ_p is the positive-sequence impedance angle of the three-phase induction machine.

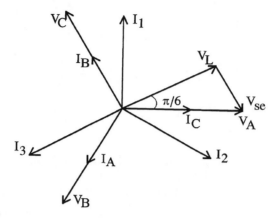

Figure 5.29 *Phasor diagram of single-phase SRSEIG when the three-phase machine is balanced at a unity-power-factor load*

Figure 5.29 shows the phasor diagram of the SRSEIG under perfect phase balance conditions. From the diagram, the following relationships can be deduced:

$$|Z_{se}| = X_{se} = \frac{Z_p}{2\sqrt{3}} \tag{5.35}$$

$$\frac{R_L}{a} = \sqrt{3}X_{se} = \frac{|Z_p|}{2} \tag{5.36}$$

and

$$V_L = \frac{\sqrt{3}}{2}V_A. \tag{5.37}$$

From (5.33) and (5.35),

$$X_{sh} = 2X_{se} \tag{5.38}$$

or

$$C_{se} = 2C_{sh}. \tag{5.39}$$

Equations (5.36), (5.38) and (5.39) imply that perfect phase balance in the three-phase induction machine can be achieved when a compensation factor of 0.5 is used and the load resistance (referred to the base frequency) is numerically equal to one-half of the positive-sequence impedance. Under these conditions, the load voltage is 0.866 times the terminal voltage of the three-phase machine, while the generator impedance angle is equal to $2\pi/3$ electrical radians.

Capacitances for perfect phase balance. A practical design problem is the selection of C_{sh} and C_{se} to give perfect phase balance for a given load resistance R_L. When the speed of the IG is regulated to give a constant frequency in the output voltage, C_{sh} and C_{se} can be determined directly from (5.33) and (5.35). For constant speed operation, however, the per-unit frequency is a function of the load and the excitation capacitances, which are not known until the SRSEIG circuit is completely solved. To overcome this difficulty, the following iterative solution procedure is proposed:

1. Input the values of per-unit speed and load resistance R_L.
2. Assume an initial value of the per-unit frequency a.
3. Compute X_{sh} and X_{se} from (5.33) and (5.35).
4. Solve the SRSEIG circuit for a, X_m and Z_p using the above values of X_{se} and X_{sh}.
5. Repeat steps 3 and 4 until the per-unit frequency a in successive iterations differs by a sufficiently small value (say 1.0e 6).
6. Determine C_{sh} and C_{se} using the final values of a and X_m.
7. Compute the generator performance.

Figure 5.30 shows the computed and experimental values of the shunt and series capacitances to give balanced operation for the experimental machine. The values of C_{sh} and C_{se} required increase approximately linearly with the load admittance. At light loads, the combination of R_L, C_{sh} and C_{se} may force the magnetizing reactance of the IG to exceed the unsaturated value, implying that the operating point does not exist.

The results in Figure 5.30 show that, by a proper selection of the excitation capacitance, compensation capacitance and load resistance, it is possible to obtain perfect balance in the three-phase machine over a wide range of load.

Voltage unbalance factor. Balanced operation can only be achieved for a given combination of R_L, C_{sh} and C_{se}. When the load or the compensation factor changes, the three-phase machine will again be unbalanced. The degree of unbalance is

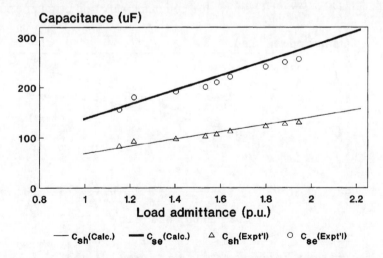

Figure 5.30 *Computed and experimental values of capacitances to give balanced operation in the three-phase machine at unity-power-factor loads*

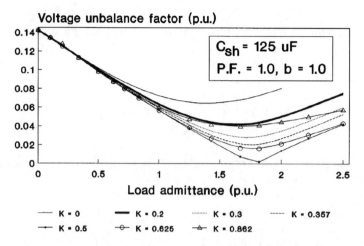

Figure 5.31   *Variation of voltage unbalance factor with load admittance*

conveniently described in terms of the voltage unbalance factor (VUF) which is defined as the ratio of the negative-sequence voltage V_n to the positive-sequence voltage V_p. Figure 5.31 shows the effect of compensation factor K on the VUF when the shunt excitation capacitance is constant at $125\,\mu$F. It is interesting to note that for each value of K, the variation of VUF with load admittance is a V-shaped curve. The VUF is smaller, the closer the value of K is to 0.5. When K is equal to 0.5, the minimum phase imbalance in the three-phase machine is obtained, with zero VUF occurring at a load admittance of 1.82 p.u.

Output power and efficiency.   For a given value of shunt excitation capacitance, the output power from the single-phase SRSEIG depends on the series compensation capacitance used. Figure 5.32 shows the computed variation of output power of

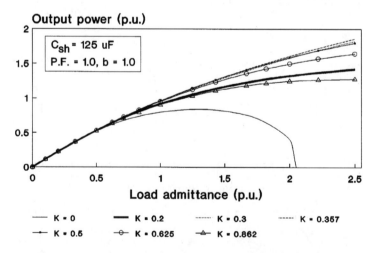

Figure 5.32   *Variation of output power with load admittance*

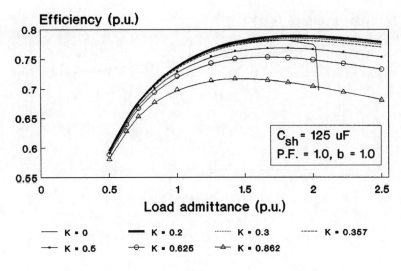

Figure 5.33   *Variation of efficiency with load admittance*

the experimental machine with load admittance when C_{sh} is equal to 125 μF. With no compensation (i.e. $K = 0$), a power limit occurs at a load admittance of about 1.3 p.u and de-excitation occurs when the load admittance is equal to 2.05 p.u. When K exceeds 0.2, a marked increase in the output power is obtained and stable operation of the single-phase SRSEIG is possible even with very large values of load admittance. The output power from the machine is large when K is between 0.3 and 0.5, the best power output being obtained when K is equal to 0.357. This observation suggests that the use of very large values of compensation capacitance is unnecessary and should be avoided.

Figure 5.33 shows the computed variation of efficiency of the single-phase SR-SEIG with load admittance for various values of K. The efficiency is highest when K is equal to 0.2 and it decreases monotonously as K increases. The efficiency of the uncompensated machine is comparable with that of the SRSEIG when K is between 0.2 and 0.357, up to a load admittance of 2.0 p.u. at which it drops abruptly due to the rapid decrease in the output voltage.

5.4.3   Experimental Results and Discussion

In order to validate the performance analysis of the single-phase SRSEIG, load tests were performed on the experimental machine IG1. Attention was focused on the constant speed mode of operation and resistive loads. Figure 5.34 shows the computed and experimental load characteristics of the single-phase SRSEIG for various values of K, with C_{sh} constant at 125 μF. Without series compensation, the load characteristic of the single-phase SEIG has the familiar 'bend' and the maximum current that can be supplied is 1.14 p.u., at which the voltage drop is approximately 40 %. With K equal to 0.5, the load characteristic of the single-phase

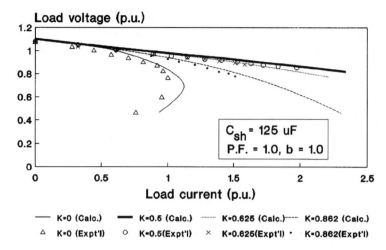

Figure 5.34   *Computed and experimental load characteristics of single-phase SRSEIG ($C_{sh} =$ 125 μF)*

SRSEIG is practically linear up to a load current of 2.0 p.u. at which the voltage drop is 22 %.

Figure 5.35 shows the load characteristics of the single-phase SRSEIG for various values of K, with C_{se} at 250 μF. The results indicate that, by varying the shunt excitation capacitance, the load voltage can be regulated. By selecting proper values of C_{sh}, it is possible to obtain different degrees of compounding in the load characteristic. When K is equal to 0.38, for example, a nearly level compounded characteristic is obtained, with zero voltage regulation occurring at a load current of 1.3 p.u. A reduction of K (or C_{sh}) results in a lower output voltage as well as a

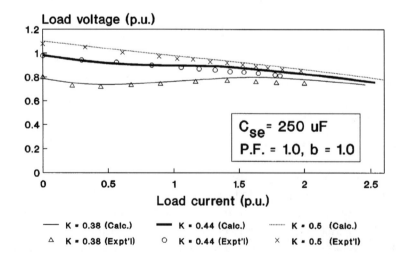

Figure 5.35   *Computed and experimental load characteristics of single-phase SRSEIG*

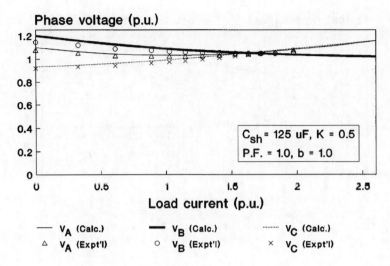

Figure 5.36 *Computed and experimental phase voltage characteristics of single-phase SRSEIG*

reduction in VUF, but perfect phase balance cannot be achieved when K deviates from 0.5.

Figures 5.36–5.38 show the computed and experimental characteristics of the single-phase SRSEIG when C_{sh} is constant at 125 μF and K is equal to 0.5. From Figure 5.36 and Figure 5.37, it is observed that the three-phase machine is balanced at a load current of 1.6 p.u., at which the phase voltage and current are 1.05 p.u. and 0.94 p.u., respectively. The A-phase voltage varies only slightly with

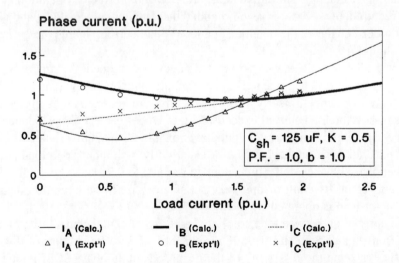

Figure 5.37 *Computed and experimental phase current characteristics of single-phase SRSEIG*

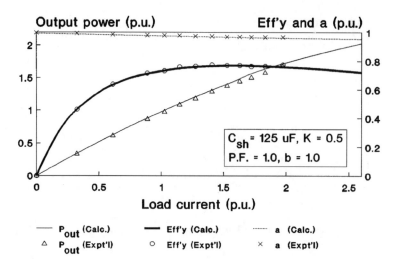

Figure 5.38 *Computed and experimental output power, efficiency and frequency characteristics of single-phase SRSEIG*

load current, exhibiting a concave-upward characteristic. On the other hand, the B-phase voltage decreases and the C-phase voltage increases monotonously with load current. Figure 5.37 also shows that, under light-load conditions, overcurrent occurs in B-phase. When the load current exceeds that corresponding to perfect phase balance, overcurrent will first occur in A-phase and C-phase. At very large load currents, all three phases of the induction generator will be overloaded. Provided that the load current varies between 1.0 and 2.0 p.u., the phase imbalance in the three-phase machine should be acceptable. When prolonged operation at light load is required, however, it is recommended that smaller values of C_{sh} and C_{se} be used in order to balance the machine at a lower load current, thereby reducing the VUF at light load.

Figure 5.38 shows the output power, efficiency and frequency characteristics of the single-phase SRSEIG. Due to the self-regulation in the load voltage, the power output increases almost linearly as the load current increases. A high efficiency is obtained when the load current exceeds 0.8 p.u. Under perfect phase balance conditions, the machine delivers an output power of 1.48 p.u. (1760 W) at an efficiency of 0.77 p.u. The per-unit frequency a drops slightly with increase in load current: at perfect phase balance, the p.u. frequency is equal to 0.964.

The above results confirm that very satisfactory machine operation at unity-power-factor load is obtained when $K = 0.5$.

Good correlation between the computed and experimental characteristics is observed from Figure 5.30 and from Figures 5.34 to 5.38. The validity of the theoretical analysis and the feasibility of the proposed single-phase SRSEIG are thus verified.

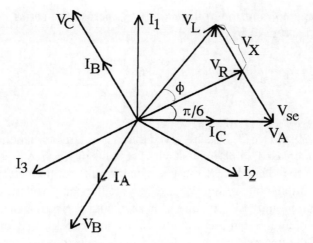

Figure 5.39 *Phasor diagram of single-phase SRSEIG when the three-phase machine is balanced at a lagging-power-factor load*

5.4.4 Effect of Load Power Factor

Autonomous power systems typically supply lighting and storage heating loads with power factors equal to or close to unity. For these applications, the analysis and results presented in Sections 5.4.2 and 5.4.3 are generally applicable. Special consideration should be given, however, to the case when the load power factor differs from unity.

Figure 5.39 shows the phasor diagram of the single-phase SRSEIG under perfect phase balance and the load power factor angle is ϕ lagging. From the diagram, the following relationships can readily be deduced:

$$R_L = \frac{|Z_p|}{\sqrt{3}} \cdot \cos\frac{\pi}{6} \tag{5.40}$$

$$X_L = \frac{|Z_p|}{\sqrt{3}} \cdot \cos\frac{\pi}{6} \cdot \tan\phi \tag{5.41}$$

$$X_{se} = \frac{|Z_p|}{\sqrt{3}} \cdot \frac{\sin(\phi + \pi/6)}{\cos\phi}. \tag{5.42}$$

The compensation factor K that results in perfect phase balance in the three-phase IG is

$$K = \frac{X_{se}}{X_{sh}} = \frac{\sin(\phi + \pi/6)}{\cos\phi}. \tag{5.43}$$

The load voltage V_L and the voltage drop V_{se} across the series compensation capacitance C_{se} are given by

$$V_L = V_{ph}. \cos\frac{\pi}{6}. \sec\phi \tag{5.44}$$

$$V_{se} = V_{ph}.\frac{\sin(\phi + \pi/6)}{\cos\phi}. \tag{5.45}$$

From Figure 5.39 and (5.43), it is observed that X_{se} is larger when the load power factor angle is increased, implying that a higher value of K is required. At a power factor angle of $\pi/6$ rad lagging, K must be equal to unity for perfect phase balance. Under this condition, both V_L and V_{se} are equal to the phase voltage of the IG.

A smooth variation of K is required for achieving perfect balance at different load power factor angles, but this would significantly increase the complexity and capital cost of the generator system. In practice, provision of discrete values of C_{se} and K suffices to give satisfactory generator performance. Figure 5.40 shows the load characteristics of the single-phase SRSEIG at different power factors when $C_{sh} = 125\mu F$ and $K = 1.0$. It is observed that the load voltage exhibits an interesting double-peak characteristic when the load power factor is lagging. The undulation in the load voltage characteristics is larger as the load power factor decreases, but the load voltage remains substantially constant at light and medium loads. At a power factor of 0.8 lagging, zero voltage regulation occurs at three different values of load current, but the computed results indicate that for load currents not exceeding 1.5 p.u., the voltage regulation is smallest when the load power factor is 0.866 lagging. With unity-power-factor loads, however, the voltage regulation is much larger compared with that obtaining when a compensation factor of 0.5 is used (Figure 5.34).

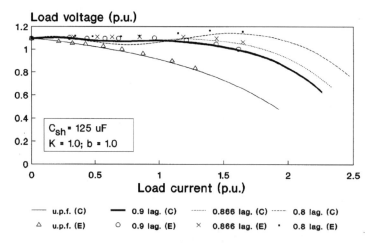

Figure 5.40 *Computed and experimental load voltage characteristics of single-phase SRSEIG at different load power factors*

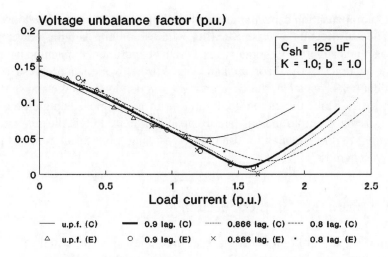

Figure 5.41 *Computed and experimental variations of VUF of single-phase SRSEIG at different load power factors*

Figure 5.41 shows the computed and experimental VUF characteristics of the single-phase SRSEIG at different power factors when $K = 1.0$. At a load power factor of 0.866 lagging, perfect phase balance is obtained at a load current of 1.65 p.u., which is very close to the rated value. At a load power factor of 0.8 lagging and 0.9 lagging, the minimum values of VUF are 0.018 and 0.01, respectively. At unity-power-factor load, the VUF for heavy loads is considerably higher.

For practical applications, the value of K may be changed by a simple switching operation on C_{se} which comprises two identical units, each equal to C_{sh}. For load power factors close to unity, the two capacitor units are connected in parallel to yield a compensation factor of 0.5. As the load power factor becomes more lagging, one of the capacitor units may be switched out to yield a compensation factor of unity.

5.4.5 Summary

In this section, the steady-state performance of a single-phase, self-regulated SEIG using a three-phase machine has been analysed. The effect of compensation factor on the generator performance has been investigated in detail with reference to a small experimental machine. Very good phase balance is obtained over a wide range of load by virtue of the Steinmetz connection and the phase shifting effect of the series compensation capacitance on the load current. Other advantages of the single-phase SRSEIG include good winding utilization, large power output, high efficiency and small voltage regulation. The conditions for achieving perfect phase balance in the three-phase machine, which supplies single-phase loads, have also been deduced from the symmetrical components analysis and the phasor diagram. An iterative method for computing the values of shunt capacitance and series compensation capacitance for phase balancing has been developed. The performance of the single-phase SRSEIG when supplying lagging power factor loads is also investigated. Laboratory tests on the experimental machine have confirmed the accuracy of the theoretical analysis. Further work is in progress to study the transient performance of the generator, e.g. sudden load change and switching of motor loads. Since the circuit configuration of the proposed single-phase SR-SEIG is extremely simple and only static capacitors are required, the generator can be conveniently implemented for use in low-cost single-phase autonomous power generation schemes.

5.5   SEIG with the Smith Connection

Application of the Smith connection to a three-phase IG connected to a single-phase grid has been studied in Section 3.3. In this section, it will be shown that this connection can be developed to give a novel excitation scheme for a three-phase SEIG that supplies isolated single-phase loads. Good phase balance in the generator can be achieved, resulting in a high efficiency and a large power output.

5.5.1   Circuit Connection and Operating Principle

Figure 5.42 shows the proposed excitation scheme. When viewed at terminals 1 and 3 across which the single-phase load Z_L is connected, the stator phases and the excitation capacitances are in the form of the Smith connection discussed in Section 3.3. For easy reference in the subsequent discussion, this new excitation

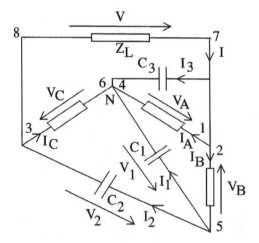

Figure 5.42   *Circuit connection of proposed excitation scheme for three-phase IG supplying an isolated single-phase load*

scheme will be abbreviated as the SMSEIG. In order to suppress the zero-sequence current, the capacitance C_3 must be equal to twice the capacitance C_2.

Self-excitation in the SMSEIG takes place in a similar manner as a three-phase SEIG with symmetrical winding connection and balanced capacitances. Residual flux must be present in the rotor, and the capacitances must be sufficiently large for the terminal voltage to build up [2]. As in other SEIG schemes, the frequency and magnetizing reactance of the SMSEIG are not constant but vary with the rotor speed and the load impedance.

The phasor diagrams for the SMSEIG are the same as that for the SMIG, as illustrated in Figures 3.13(a) and (b). Provided that the generator impedance angle ϕ_p lies between $2\pi/3$ rad and $5\pi/6$ rad, I_B can be synthesized with the proper magnitude and phase angle so as to yield perfect phase balance in the SMSEIG. Under this condition the IG operates with balanced phase currents and phase voltages and its performance is similar to a three-phase SEIG with balanced excitation capacitances and balanced load impedances. The currents I_1, I_2 and I_3 can be adjusted easily by varying the capacitances C_1, C_2 and C_3. Figure 3.12 also suggests that the SMSEIG is best suited for supplying high-power-factor (e.g. resistive) loads.

5.5.2 Performance Analysis

To determine the steady-state performance of the SMSEIG, the analysis presented in Section 3.3 for the grid-connected SMIG is used in association with the solution method for a single-phase SEIG, presented in Section 5.2.

For the SMSEIG, it can be shown that the input impedance Z_{in} across terminals 1 and 3 is given by

$$Z_{in} = R_{in} + jX_{in} = \frac{Y_p + Y_n + 3Y_1 + 2Y_2}{2Y_1Y_2 + (Y_p + Y_n)(Y_1 + Y_2) + Y_pY_n} \tag{5.46}$$

where $Y_1 = ja^2.2\pi f_{base}C_1$, $Y_2 = ja^2.2\pi f_{base}C_2$ and $Y_3 = ja^2.2\pi f_{base}C_3$.

For a given per-unit speed b and a given set of excitation capacitances, the values of a and X_m may be determined by minimizing the following scalar impedance function:

$$Z(a, X_m) = \sqrt{\left(\frac{R_L}{a} + R_{in}\right)^2 + (X_L + X_{in})^2}. \tag{5.47}$$

After a and X_m have been determined, the steady-state performance of the SMSEIG can be obtained from the appropriate circuit equations.

5.5.3   Balanced Operation

Conditions for phase balance. As in the SMIG, the capacitive admittances for perfect phase balance are given by

$$|Y_1| = \frac{2}{\sqrt{3}}|Y_p| \sin\left(\frac{5\pi}{6} - \phi_p\right) \tag{5.48}$$

$$|Y_2| = |Y_p| \sin\left(\phi_p - \frac{2\pi}{3}\right) \tag{5.49}$$

$$|Y_3| = 2|Y_p| \sin\left(\phi_p - \frac{2\pi}{3}\right) \tag{5.50}$$

while the load admittance Y_L, referred to the base frequency, is given by

$$|Y_L| = -\frac{|I|}{|V|} = \frac{|Y_p|}{\sqrt{3}}\sqrt{1 + 8\sin^2\left(\phi_p - \frac{2\pi}{3}\right)}. \tag{5.51}$$

Equations (5.48)–(5.50) may be simplified when the load is purely resistive. Since the load power factor angle is now π rad, it can be shown that the generator impedance angle ϕ_p is

$$\phi_p = \tan^{-1}\left(-\frac{2}{\sqrt{3}}\right) = 2.2845 \text{ rad.} \tag{5.52}$$

The IG thus operates with an output power factor of 0.655 leading when perfectly balanced and supplying a pure resistive load.

Substitution of (5.52) into (5.48) to (5.50) yields

$$|Y_1| = \frac{|Y_2|}{2} = |Y_3| = \frac{1}{\sqrt{7}}|Y_p|. \tag{5.53}$$

The capacitances C_1 and C_3 are thus equal, while the currents are related by

$$|I| = \sqrt{3}|I_1| = 3|I_2| = 3|I_3| = \frac{3}{\sqrt{7}}I_{ph}. \tag{5.54}$$

Capacitances for perfect phase balance. It is of interest to determine, for a given load impedance and per-unit speed, the values of excitation capacitances that give perfect phase balance in the SMSEIG. Since Y_p and ϕ_p are functions of a and X_m, both being unknown to start with, an iterative procedure has to be used. The steps are outlined as follows:

1. Assume appropriate initial values of a and X_m.
2. For the given value of load impedance, compute the load admittance $|Y_L|$ and the operating load power factor angle ϕ.
3. Determine the corresponding generator impedance angle ϕ_p using an iterative method.

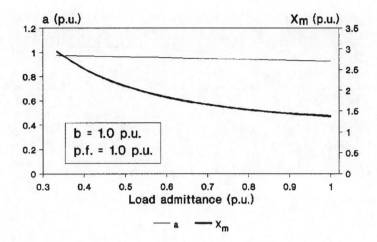

Figure 5.43 *Computed variation of per-unit frequency and magnetizing reactance under perfect phase balance*

4. Compute $|Y_p|$ from (5.51), using the current values of $|Y_L|$ and ϕ_p.
5. Compute the capacitive admittances $|Y_1|$, $|Y_2|$, and $|Y_3|$ from (5.48)–(5.50).
6. Determine the new values of a and X_m, using the solution technique outlined in Section 5.2.2.
7. Update the values of ϕ and $|Y_L|$ using the new values of a and X_m.
8. Repeat steps 3 to 7 until the values of a in successive iterations are less than the prescribed tolerance, say 1.0e-6.
9. Compute the excitation capacitances C_1, C_2 and C_3 using the final values of a and X_m, hence obtain the performance of the SMSEIG under balanced conditions.

5.5.4 Results and Discussion

Computed results: constant speed operation. Computed results were obtained for experimental machine IG1 whose data are given in Appendix D.1, the emphasis being placed on the performance when the SMSEIG is operated at constant speed and is supplying unity-power-factor loads.

Figure 5.43 shows the computed variations of a and X_m with load admittance Y_L when the SMSEIG operates with perfect phase balance. A reduction in Y_L causes X_m to increase, implying that the induction machine becomes less saturated. When $Y_L = 0.403$ p.u., X_m is equal to the unsaturated value X_{mu} ($= 2.48$ p.u.). There is thus a value of load admittance below which perfect phase balance cannot be achieved for the SMSEIG.

Figure 5.44 shows the variation of C_1 with Y_L to give perfect phase balance and the corresponding phase voltage of the induction generator. The capacitance required for giving perfect phase balance decreases as Y_L is reduced, hence the decrease in X_m.

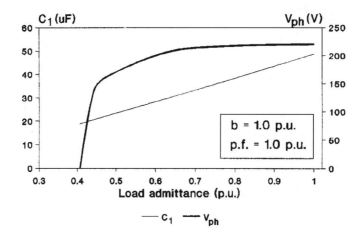

Figure 5.44 *Computed variation of capacitance C_1 and phase voltage under perfect phase balance*

Experimental results: constant speed operation. To verify the phase balancing capability of the SMSEIG, laboratory tests were performed on the experimental machine. The speed of the SMSEIG was maintained at rated value (i.e. $b = 1.0$ p.u.) and the capacitances were increased until self-excitation occurred. The load resistance was then switched in and the capacitances were adjusted until the phase voltages and currents were balanced. Since the load was resistive, the balance point could be obtained quite easily by varying the capacitances simultaneously in the proportion as prescribed by (5.53). Both machine noise and vibration levels were low when the SMSEIG was operating under perfect phase balance. The performance characteristics are given in Figures 5.45–5.47.

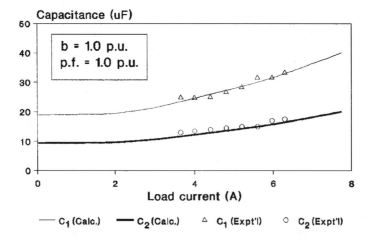

Figure 5.45 *Excitation capacitances to give perfect phase balance in SMSEIG*

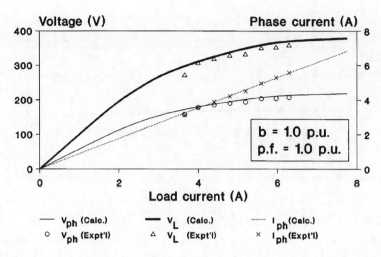

Figure 5.46 *Variation of load voltage, phase voltage and phase current of SMSEIG under perfect balance*

Figure 5.45 shows the values of C_1 and C_2 that result in perfect phase balance. The good correlation between the computed and experimental results confirms the feasibility of the proposed excitation scheme. It is observed that larger values of capacitances need to be used for achieving perfect phase balance at heavier loads.

Figure 5.46 shows the corresponding variations of load voltage, generator phase voltage and generator phase current with load current. Due to the increased reactive power from the capacitances, the generator is driven into heavier saturation and the output voltage increases. When delivering a rated load current of 6.2 A, the

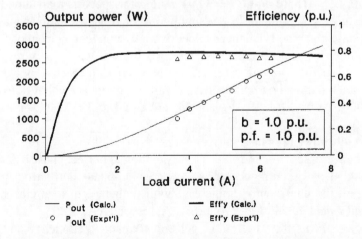

Figure 5.47 *Output power and efficiency characteristics of SMSEIG under perfect phase balance*

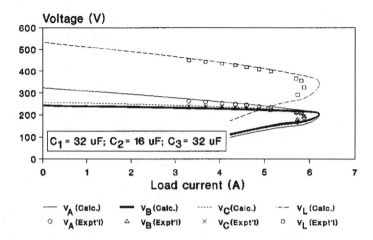

Figure 5.48   *Phase voltage and load voltage characteristics of SMSEIG with constant excitation capacitances*

phase voltage is 213 V (0.97 p.u.). This shows that the generator rating has been fully utilized.

Figure 5.47 shows the output power and efficiency characteristics of the SMSEIG under balanced conditions. At rated load current, the generator delivers a power of 2250 W to the single-phase load, which is approximately equal to the rated output when the machine is run as a three-phase motor. The generator efficiency is high since only positive-sequence losses are incurred. For load currents from 1.7 A to rated value, the efficiency exceeds 0.75 p.u. and remains substantially constant. The maximum efficiency is 0.79 and occurs at a load current of 3.5 A.

Figures 5.48–5.50 show the performance characteristics of the SMSEIG when $C_1 = 32\,\mu F$, $C_2 = 16\,\mu F$ and $C_3 = 32\,\mu F$. The rotor speed is maintained at rated value and a unity-power-factor load is supplied. With these excitation capacitances, the three-phase machine is balanced experimentally at a load current of 5.8 A, the corresponding phase voltage and phase current being 203 V and 5.2 A, respectively. Since the capacitances remain unchanged, the excitation reactive power will be excessive when the load current decreases, causing the phase voltages to increase (Figure 5.48). At no load, V_A, V_B and V_C are equal to 324 V, 243 V and 256 V, respectively.

Figure 5.49 shows the phase current characteristics of the SMSEIG. When the load current decreases from the value corresponding to perfect phase balance, both A-phase and B-phase currents increase, whereas C-phase current decreases. At light load, all the phase currents increase with reduction in load current. At no load, I_A and I_B are equal to 11.2 A and 8.1 A, respectively.

Figure 5.50 shows the variation of per-unit frequency, efficiency and VUF of the SMSEIG with load current. For operation at rated speed, the output frequency is 0.95 p.u. at rated current. The efficiency drops significantly with reduction of

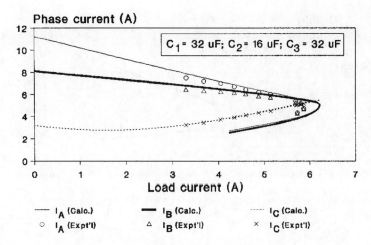

Figure 5.49   *Phase current characteristics of SMSEIG with constant excitation capacitances*

load current due to the increase in the negative-sequence voltage and current components, as well as the increase in iron loss due to heavier magnetic saturation. Variation in the VUF is approximately linear from no load to the balance point.

The validity of the analysis and solution method is verified by the close agreement between the computed and experimental results in Figures 5.45–5.50.

Computed results: variable speed operation. Pertinent characteristics of the SMSEIG were also computed for variable speed operation. Figure 5.51 shows the computed values of C_1 required for balanced operation of the SMSEIG at different operating speeds and the corresponding values of the phase voltage. For a given load impedance, C_1 (and also C_2 and C_3) decreases with increase in the

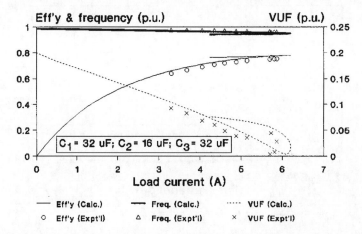

Figure 5.50   *Efficiency, frequency and VUF characteristics of SMSEIG with constant excitation capacitances*

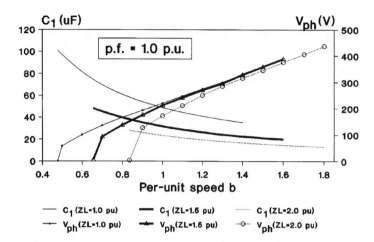

Figure 5.51   *Computed variations of C_1 and phase voltage of SMSEIG with rotor speed under perfect phase balance*

per-unit speed b. There is also a minimum value of b below which stable operation is not possible while maintaining perfect balance. The phase voltage (and hence the load voltage) increases with b, thereby imposing an upper limit to the permissible operating speed.

Figure 5.52 shows the computed performance of the SMSEIG as a function of speed when $C_1 = 32\,\mu\text{F}$, $C_2 = 16\,\mu\text{F}$ and $C_3 = 32\,\mu\text{F}$, and a constant load impedance of 1.48 p.u. at unity power factor is being supplied. This combination of excitation capacitances and load impedance results in perfect phase balance

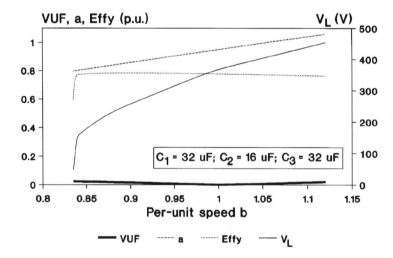

Figure 5.52   *Computed variations of VUF, per-unit frequency, efficiency and load voltage of SMSEIG with rotor speed*

at rated speed. The VUF characteristic is a nearly flat V-shaped curve, implying that the phase imbalance is very slight despite a large change in the rotor speed. Over a wide speed range, the efficiency of the SMSEIG remains practically constant at 0.78 p.u., while the per-unit frequency a varies almost linearly with speed. The load voltage V_L decreases with reduction in rotor speed and it collapses at $b = 0.83$ p.u. The above results indicate that, once the SMSEIG has been balanced for a given load impedance at rated speed, its performance at other speeds will also be satisfactory.

5.5.5 Summary

Section 5.5 has presented the principle and analysis of a novel excitation scheme for a stand-alone three-phase IG that supplies single-phase loads, namely the SMSEIG. By adopting the Smith connection with appropriate values of excitation capacitances, balanced operation of the three-phase machine can be achieved. The steady-state performance of the SMSEIG is analysed using the method of symmetrical components in association with an optimization procedure. A method to determine the capacitances to give perfect phase balance is also presented. Experimental investigations have confirmed the feasibility of the proposed excitation scheme. Although only results of the resistive load case have been reported, the analysis set forth can readily be applied to different load power factor conditions. The SMSEIG has the advantages of low cost, high efficiency and large power output, and as such is an economical choice when developing autonomous single-phase power systems in remote regions.

References

[1] T.F. Chan, 'Performance analysis of a three-phase induction generator self-excited with a single capacitance," *IEEE Transactions on Energy Conversion*, Vol. 14, No. 4, pp. 894–900, December 1999.

[2] S.S. Murthy, O.P. Malik and A.K. Tandon, 'Analysis of self-excited induction generators," *IEE Proceedings*, PC, Vol. 129, No. 6, pp. 260–265, November 1982.

[3] L. Ouazene and G. McPherson, Jr, 'Analysis of the isolated induction generator," *IEEE Transactions on Power Apparatus and Systems*, Vol. PAS-102, No. 8, pp. 2793–2798, August 1983.

[4] Byron S. Gottfried and Joel Weisman, *Introduction to Optimization Theory*, Prentice Hall, Englewood Cliffs, NJ, 1973.

[5] A.I. Alolah and M.A. Alkanhal, 'Optimization-based steady state analysis of three-phase self-excited induction generator,' *IEEE Transactions on Energy Conversion*, Vol. 15, No. 1, pp. 61–65, March 2000.

[6] A.K. Jabri and A.I. Alolah, 'Capacitance requirement for isolated self-excited induction generator,' *IEE Proceedings*, PB, Vol. 137, No. 3, pp. 154–159, May 1990.

[7] T.F. Chan, 'Capacitance requirements of self-excited induction generator,' *IEEE Transactions on Energy Conversion*, Vol. 8, No. 2, pp. 304–311, June 1993.

[8] A. Balfour and W.T. Beveridge, *Basic Numerical Analysis with Fortran*, Heinemann, London, pp. 138–144, 1977.

[9] R. Holland, 'Appropriate technology - rural electrification in developing countries,' *IEE Review*, Vol. 35, No. 7, pp. 251–254, August 1989.

[10] A.I. Alolah and M.A. Alkanhal, 'Excitation requirements of three phase self excited induction generator under single phase loading with minimum unbalance,' *Proceedings of IEEE Power Engineering Society 2000 Winter Meeting*, 23–27 January 2000, Singapore.

6

Voltage and Frequency Control of SEIG with Slip-Ring Rotor

6.1 Introduction

A major disadvantage of an SEIG is that the output voltage and frequency are speed and load dependent. An increase in the rotor speed, for example, will result in a proportionate increase in frequency, often accompanied by severe overvoltage and excessive current. There has been rigorous research on the voltage and frequency control of squirrel-cage-type SEIGs [1–6], but relatively little research effort has been devoted to the use of the slip-ring induction machine for generator applications. Although the slip-ring machine is more expensive and requires more maintenance, it permits rotor slip-power control when driven by a variable speed turbine. The slip-ring machine may be operated as a double-output induction generator (DOIG) using the slip energy recovery technique [7, 8]. In the case of a self-excited slip-ring induction generator (SESRIG), the system cost can be further reduced by the use of a simple rotor resistance controller [9, 10]. Since only a capacitor bank need be connected to the stator terminals, the SESRIG provides a good-quality AC source with little harmonic distortion to the stator load. Another advantageous feature of the SESRIG is that independent control of the voltage and frequency can be easily achieved. Even with a wide variation in speed, the generator frequency can be maintained reasonably constant by rotor resistance control, while the voltage can be controlled by varying the excitation capacitance. The rating of the rotor resistance controller is small compared with the generator rating, hence the cost saving is quite significant.

In this chapter, the voltage and frequency control of a three-phase SESRIG by variation of external rotor resistance will be investigated. Based on a normalized

Distributed Generation: Induction and Permanent Magnet Generators L. L. Lai and T. F. Chan
© 2007 John Wiley & Sons, Ltd

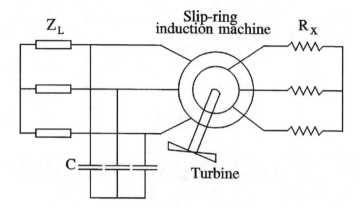

Figure 6.1   *Self-excited slip-ring induction generator (SESRIG)*

equivalent circuit model, the frequency and voltage control characteristics are deduced and experimental results are presented to verify the feasibility of the control method. Practical implementation of a closed-loop scheme that uses chopper-
controlled rotor resistance will also be described.

6.2   Performance Analysis of SESRIG

Figure 6.1 shows the circuit arrangement of a three-phase SESRIG which is supplying a balanced stator load. The excitation capacitance C is required for initiating
voltage build-up and maintaining the output voltage. Note that the electric output
power is dissipated in both the stator impedance Z_L and the external rotor resistance R_x, hence the machine may also be regarded as a DOIG if the power in R_x
is effectively utilized.

Figure 6.2 shows the per-phase equivalent circuit of the SESRIG, where the
rotor resistance R_2 is the sum of the rotor winding resistance and the external rotor
resistance, both referred to the stator side.

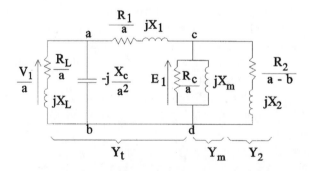

Figure 6.2   *Per-phase equivalent circuit of SESRIG*

Various methods have been developed for solution of the SEIG equivalent circuit. Adopting the nodal admittance method [11], the following relationship may be established for successful voltage build-up:

$$Y_t + Y_m + Y_2 = 0 \tag{6.1}$$

where

$$Y_t = \frac{1}{Z_t} = \frac{1}{Z_{ac} + Z_{ab}} = G_t + jB_t \tag{6.2}$$

$$Y_m = \frac{a}{R_c} - j \cdot \frac{1}{X_m} = G_m - jB_m \tag{6.3}$$

$$Y_2 = \frac{1}{\frac{R_2}{a-b} + jX_2} = G_2 - jB_2. \tag{6.4}$$

Equating the real and imaginary parts in (6.1) to zero respectively, the following equations in real numbers are obtained:

$$G_t + G_m + G_2 = 0 \tag{6.5}$$

$$B_t - B_m - B_2 = 0. \tag{6.6}$$

For a given rotor speed, load impedance and excitation capacitance, (6.5) is a nonlinear equation in the variable a only. Numerical solution of (6.5) using an iterative method [12] enables a to be determined, and (6.6) can subsequently be used to calculate X_m. With the aid of the magnetization curve (plot of E_1 versus X_m), E_1 can be determined and the equivalent circuit is completely solved to yield the steady-state performance.

Performance analysis and experiments for variable speed operation were conducted on a three-phase, four-pole, 50 Hz, 380 V, 4.5 A, 1.8 kW, star/star-connected slip-ring induction machine whose per-unit equivalent circuit constants are: $R_1 = 0.0597$, $X_1 = 0.118$, $R_2 = 0.0982$, $X_2 = 0.118$. The magnetization curve was represented by the following set of descriptive equations:

$$E_1 = \begin{cases} 1.4613 - 0.3327X_m, & X_m < 1.7728 \\ 1.5294 - 0.3711X_m, & 1.7728 \leq X_m < 2.045 \\ 3.0455 - 1.1125X_m, & 2.045 \leq X_m < 2.213 \\ 185.1 \ - 83.37X_m, & 2.213 \leq X_m < 2.22 \\ 0, & 2.22 \leq X_m. \end{cases} \tag{6.7}$$

Figures 6.3, 6.4 and 6.5 show, respectively, the stator voltage, stator current and frequency characteristics of the SESRIG for different values of external rotor resistance R_x. For convenience, all the machine parameters, except the excitation capacitance, are expressed per unit. It is observed that increasing R_x has the effect of shifting the performance characteristics to the right-hand side of the speed axis. At a rotor speed of 1.05 p.u. or above, the generator voltage or frequency can be

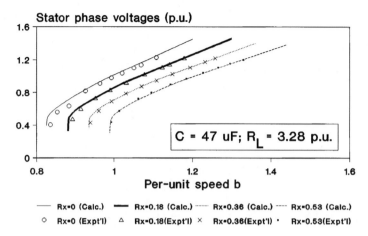

Figure 6.3 *Stator voltage variation of SESRIG with rotor speed at different values of external rotor resistance*

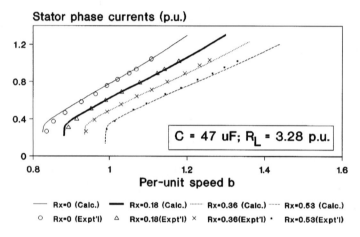

Figure 6.4 *Stator current variation of SESRIG with rotor speed at different values of external rotor resistance*

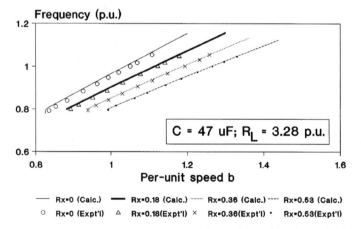

Figure 6.5 *Frequency variation of SESRIG with rotor speed at different values of external rotor resistance.*

maintained at rated value (i.e. 1.0 p.u.) by varying R_x. This feature will be employed for voltage and frequency control of the SESRIG, to be discussed in the next section.

To achieve higher system efficiency, it is important that the power dissipated in R_x be fully utilized. If R_x takes the form of resistive heater elements, the slip power could conveniently be used for storage heating, which is a common load in an autonomous power system. The total power output of the SESRIG is then the sum of the stator load power and the power consumed by R_x.

The operating speed range of the SESRIG depends on the maximum value of R_x available, the rated voltage of the rotor winding, as well as the mechanical constraints of the turbine–generator system.

6.3 Frequency and Voltage Control

In this section, the voltage and frequency control characteristics of the SESRIG will be deduced. It is assumed that both the excitation capacitance and stator load resistance remain constant, while R_x is varied with b in order to maintain a constant output frequency. For convenience, the conductance $G_e = G_t + G_m$ and the slip parameter $\gamma = a - b$ are introduced. From (6.4) and (6.5), the following equation may be written:

$$G_e + \frac{\gamma R_2}{R_2^2 + \gamma^2 X_2^2} = 0. \tag{6.8}$$

It should be noted that, for a specified value of a, G_e is a constant when the excitation capacitance and load resistance are both constant.

Solving (6.8) for γ, one obtains

$$\frac{R_2}{\gamma} = \frac{R_2}{a - b} = \frac{-1 \pm \sqrt{1 - 4G_e^2 X_2^2}}{2G_e}. \tag{6.9}$$

For practical induction generators, the term $R_2/(a - b)$ usually assumes a large negative value, hence the negative sign in the numerator of (6.9) should be chosen. Therefore,

$$\frac{R_2}{a - b} = \frac{-1 - \sqrt{1 - 4G_e^2 X_2^2}}{2G_e}. \tag{6.10}$$

Eqn. (6.10) shows that the total rotor circuit resistance should be varied linearly with the per-unit speed b in order to control the frequency at a given value.

Substituting (6.10) into (6.6),

$$\frac{1}{X_m} = B_t - \frac{2G_e^2 X_2}{1 + \sqrt{1 - 4G_e^2 X_2^2}}. \tag{6.11}$$

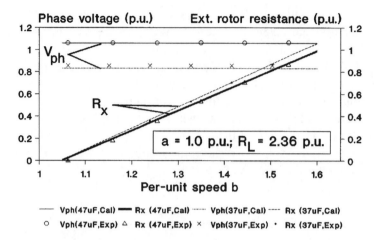

Figure 6.6 *External rotor resistance R_x for the SESRIG to operate at rated frequency and the corresponding variation of stator phase voltage*

Eqn. (6.11) implies that, for a given per-unit frequency a, excitation capacitance and load resistance, the magnetizing reactance X_m of the SESRIG, and hence the air gap voltage E_1, is independent of the rotor speed. It follows therefore that both the stator current and terminal voltage are constant.

Figure 6.6 shows the variation of external rotor resistance and the resultant output voltage when the per-unit frequency of the SESRIG is maintained at 1.0 p.u. and the stator load resistance is 2.36 p.u. Over a wide range of speed, the stator voltage remains constant at 1.06 p.u. when $C = 47\mu F$ and 0.86 p.u. when $C = 37\mu F$.

The close agreement between the computed and experimental results in Figures 6.3–6.6 confirms the accuracy of the circuit model and solution method. Figure 6.6, in addition, demonstrates the feasibility of voltage and frequency control of the SESRIG by varying the external rotor resistance. Either the frequency or the terminal voltage may be chosen as the feedback variable for voltage and frequency control.

6.4 Control with Variable Stator Load

When the stator load impedance is changed, it is also possible to maintain the output frequency constant by varying R_x, but the stator terminal voltage will differ from the nominal value. In order to control the stator terminal voltage at the desired value, it is necessary to control the excitation capacitance C simultaneously as R_x is varied. The analysis can now be formulated as the following problem:

> *For a given value of load impedance Z_L and per-unit speed b, determine the values of C and R_x that result in operation of the SESRIG at the specified voltage V_1^* and per-unit frequency a.*

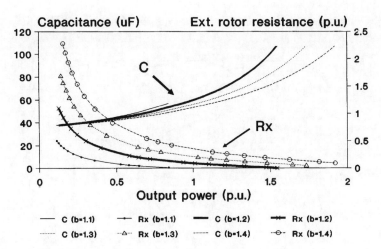

Figure 6.7 *Variation of R_x and C for the SESRIG to operate at rated voltage and rated frequency, a resistive load being supplied*

This problem can be solved by using the following simple search algorithm, with C as the variable:

1. Using the current value of search variable C, compute R_2 using (6.10).
2. Obtain the corresponding value of magnetizing reactance X_m from (6.11), and, hence, the air gap voltage E_1 from the magnetization curve.
3. Compute the stator terminal voltage V_1.
4. Evaluate the voltage difference $\Delta V = |V_1^* - V_1|$.
5. Repeat the search until ΔV is less than a specified tolerance ε, say 1.0e-6.
6. Compute the corresponding value of R_2 and, hence, R_x.

Any search algorithm based on function evaluations is suitable for the present problem, e.g. the classical golden section search method [13].

Figure 6.7 shows the computed values of C and R_x for the experimental SESRIG to operate at rated frequency and rated voltage, while supplying a stator load at a power factor of unity. At a specific rotor speed, C has to be increased when the output power increases, while R_x has to be reduced. There is thus a value of output power at which R_x is reduced to zero, which corresponds to operation with the slip rings short-circuited. At $b = 1.1$ p.u., for example, this condition prevails when the machine is delivering an output power of 0.84 p.u. When the output power exceeds this value, it is not possible to maintain the frequency and voltage simultaneously at the specified values. It is also observed from Figure 6.7 that a smaller excitation capacitance and a larger external rotor resistance are required for maintaining constant voltage and frequency at a higher rotor speed.

Figure 6.8 shows the computed variations of stator current and efficiency with output power when both the voltage and frequency of the SESRIG are maintained

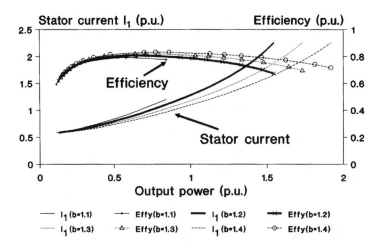

Figure 6.8 *Variation of stator current and efficiency when the SESRIG is controlled at rated voltage and rated frequency*

at rated values. At a higher rotor speed, more output power can be delivered without the rated stator current being exceeded. Because of the simultaneous control of C and R_x, the maximum efficiency of the SESRIG now occurs at a stator current of 0.91 p.u., irrespective of the rotor speed. The maximum efficiency is 0.79 p.u. when $b = 1.1$ p.u. and it increases to 0.83 p.u. when $b = 1.4$ p.u. The efficiency is good when the output power exceeds 0.4 p.u.

6.5   Practical Implementation

6.5.1   Chopper-Controlled Rotor External Resistance

It is desirable to have automatic control of the voltage and frequency when either the stator load impedance or the rotor speed changes. Instead of a variable three-phase rotor resistance, a chopper-controlled external resistance may be employed, as illustrated in Figure 6.9. Assuming that the diodes in the rotor bridge rectifier are ideal and the choke is lossless, the effective external resistance per phase R_x in the rotor circuit, referred to the stator winding, is given by [14]

$$R_x = 0.5a_t^2(1 - \alpha)R_{dc} \tag{6.12}$$

where R_{dc} = DC resistance across the chopper;
$\quad\quad\quad \alpha$ = duty cycle of the chopper;
$\quad\quad\quad a_t$ = stator/rotor turns ratio.

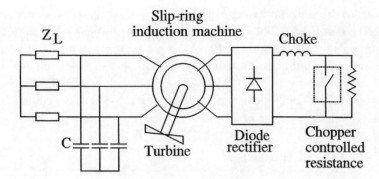

Figure 6.9 *SESRIG with chopper-controlled rotor external resistance. Reproduced by permission of T. F. Chan, K. A. Nigim and L. L. Lai, 'Voltage and frequency control of self-excited slip-ring induction generators',* Transactions on Energy Conversion, **19**, *2004: 81–87.* © *(2004) IEEE*

A reduction in the duty cycle α of the chopper results in an increase in the effective rotor resistance of the SESRIG. A variable external resistance is thus presented to the rotor circuit.

6.5.2   Closed-Loop Control

Figure 6.10 shows the block diagram for closed-loop control of the voltage and frequency of the SESRIG. The stator terminal voltage is conveniently chosen as the feedback variable since any change in speed and stator load impedance will result in a corresponding change in the terminal voltage. Referring to Figure 6.10, the stator terminal voltage signal v_t, derived from the step-down isolation transformer and signal conditioning circuit, is compared with the reference signal v_{ref} that

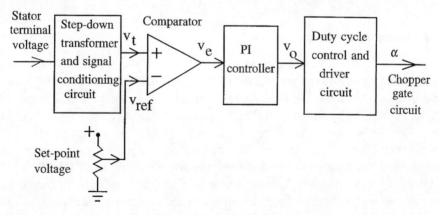

Figure 6.10 *Feedback circuit for voltage control of SESRIG. Reproduced by permission of T. F. Chan, K. A. Nigim and L. L. Lai, 'Voltage and frequency control of self-excited slip-ring induction generators',* Transactions on Energy Conversion, **19**, *2004: 81–87.* © *(2004) IEEE*

corresponds to the set-point voltage. The error signal v_e is fed to a proportional–integral (PI) controller that outputs a signal v_0 for controlling the duty cycle α of the chopper via pulse-width-modulated (PWM), opto-isolation and gate drive circuits.

In the prototype controller implemented, the chopper main switch was a power MOSFET controlled by an LM3524 PWM circuit and an IR2132 driver circuit. The total cost of the components was approximately US$50, or one-tenth that of the SESRIG system.

6.5.3   Tuning of PI Controller

Proper tuning of the PI controller is required in order to give a satisfactory dynamic performance. For this purpose, the SESRIG may be approximated as a first-order system with the following transfer function:

$$G(s) = \frac{K.e^{-st_0}}{s\tau + 1} \tag{6.13}$$

where K = system gain;
 τ = time constant of the system;
 t_0 = time delay of the system.

The parameters of the transfer function may be determined using the open-loop step response method [15]. With the transfer function identified, the gain of the proportional controller K_p and the gain of the integral controller K_i can be determined using the Ziegler–Nichols open-loop tuning method [15]. For the prototype experimental system, the controller parameters were determined as follows:

$$K = 1.15; \quad \tau = 0.8s; \quad t_0 = 0.341s;$$
$$K_p = 1.87; \quad K_i = 1.13$$

6.5.4   Dynamic Response

To study the dynamic response of the SESRIG with closed-loop control, the machine was driven by a separately excited DC motor that emulated an unregulated, variable speed turbine while a resistive load was being supplied. It was found that, with an excitation capacitance of 45 μF per phase, the terminal voltage could be maintained at the rated value over a wide speed range, the maximum rotor speed attained being limited primarily by the rated current of the DC motor. Under these conditions, the frequency of the stator voltage was found to be 48.1 Hz. Using the above value of excitation capacitance, dynamic load tests were performed on the generator system, with the reference voltage signal set to give rated stator terminal voltage. For easy comparison, the stator voltage signal from the signal conditioning circuit and the PWM control signal were monitored using a digital storage oscilloscope during the dynamic tests.

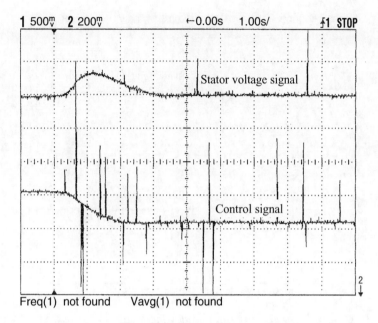

Figure 6.11 *Dynamic response of SESRIG following a speed change from 1561 r/min to 1623 r/min, $R_L = 4.29$ p.u. (time scale: 1s /div). Reproduced by permission of T. F. Chan, K. A. Nigim and L. L. Lai, 'Voltage and frequency control of self-excited slip-ring induction generators', Transactions on Energy Conversion, 19, 2004: 81–87. © (2004) IEEE*

Figure 6.11 and Figure 6.12 show the dynamic response of the SESRIG subsequent to a speed change, the stator load resistance being kept constant at 4.29 p.u. The PI controller took effect as soon as the speed started to change, outputting a corresponding PWM control signal. The stator voltage was restored to the setpoint value in approximately 2.7s when the speed was increased from 1561 r/min to 1623 r/min (Figure 6.11), and 3.2 s when the speed was decreased from 1761 r/min to 1618 r/min (Figure 6.12). The dynamic response characteristics displayed very little overshoot, showing that the controller had been properly designed with minimal overshoot and small steady-state error. It should be noted that the inherent dynamic speed–torque characteristic of the motor drive contributed partly to the delay in the voltage restoration.

Figure 6.13 shows the dynamic response of the SESRIG system after the stator load resistance was switched from 4.29 p.u. to 3.73 p.u. The load change was accompanied by a speed drop from 1645 r/min to 1590 r/min as the speed of the simulated turbine was not regulated. Again the voltage was restored in about 2.7s.

The experimental dynamic responses of the SESRIG shown in Figures 6.11–6.13 are comparable with other cage-type SEIG voltage and frequency control schemes that employ PI controllers [2, 3].

Table 6.1 shows the steady-state frequency error when the SESRIG with the feedback controller was subjected to a rotor speed change. It is observed that, despite

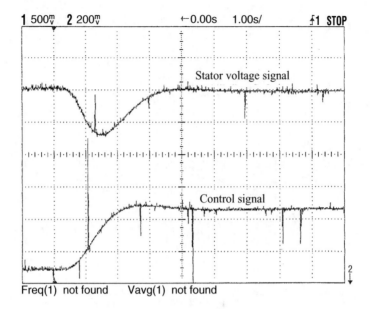

Figure 6.12 *Dynamic response of SESRIG following a speed change from 1761 r/min to 1618 r/min, $R_L = 4.29$ p.u. (time scale: 1 s/div). Reproduced by permission of T. F. Chan, K. A. Nigim and L. L. Lai, 'Voltage and frequency control of self-excited slip-ring induction generators',* Transactions on Energy Conversion, **19**, 2004: 81–87. © (2004) IEEE

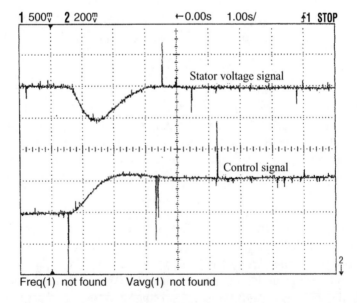

Figure 6.13 *Dynamic response of SESRIG following a change of stator load resistance from 4.29 p.u. to 3.73 p.u., accompanied by a speed change from 1645 r/min to 1590 r/min (time scale: 1 s/div). Reproduced by permission of T. F. Chan, K. A. Nigim and L. L. Lai, 'Voltage and frequency control of self-excited slip-ring induction generators',* Transactions on Energy Conversion, **19**, 2004: 81–87. © (2004) IEEE

Table 6.1 *Steady-state frequency error of SESRIG following a rotor speed change. Reproduced by permission of T. F. Chan, K. A. Nigim and L. L. Lai, 'Voltage and frequency control of self-excited slip-ring induction generators',* Transactions on Energy Conversion, *19, 2004: 81–87.* © (2004) IEEE

Stator load resistance, R_L (p.u.)	Rotor speed change (r/min)	Frequency error (Hz)
4.93	1530–1809	0.1
4.29	1561–1623	0
3.73	1588–1680	0.1

the change in rotor speed being considered, the maximum frequency deviation after the transient period is only 0.1Hz for different stator load resistances.

The above observations confirm that both the voltage and frequency can be controlled using the proposed method.

6.6 Summary

This chapter has presented the voltage and frequency control for a self-excited slip-ring induction generator (SESRIG) by varying the external rotor resistance. Steady-state performance and the control characteristics of the SESRIG have been obtained from an equivalent circuit analysis. It is shown that, with constant load impedance and excitation capacitance, both the frequency and the output voltage of the SESRIG can be maintained constant by rotor resistance control over a wide range of speed without exceeding the stator current limit. The analysis and feasibility of the control method have been verified by experiments on a 1.8 kW slip-ring machine. Practical implementation of a low-cost, closed-loop rotor resistance controller for the SESRIG has also been described. A properly tuned PI controller enables good steady-state accuracy and satisfactory dynamic response to be obtained on the generator system. The proposed scheme may be used in a low-cost, variable speed wind energy system for providing good-quality electric power to remote regions.

References

[1] E. Profumo, B. Colombo and F. Mocci, 'A frequency controller for induction generators in stand-by minihydro power plants,' *Proceedings of the 4th International Conference on Electrical Machines and Drives*, IEE Conference Publication No. 310, 13–15 September 1989, London, UK.

[2] M.A. Al-Saffa, E.-C. Nho and T.A. Lipo, 'Controlled shunt capacitor self-excited induction generator,' *Thirty-Third IEEE Industry Applications Society Annual Meeting Conference Record*, Vol. 2, pp. 1486–1490, 1998.

[3] R. Bonert and S. Rajakaruna, 'Self-excited induction generator with excellent voltage and frequency control,' *IEE Proceedings – Generation, Transmission and Distribution*, Vol. 145, No. 1, pp. 33–39, January 1998.

[4] O. Chtchetinine, 'Voltage stabilization system for induction generator in stand alone mode,' *IEEE Transactions on Energy Conversion*, Vol. 14, No. 3, pp. 298–303, September 1999.

[5] E. Suarez and G. Bortolotto, 'Voltage-frequency control of a self-excited induction generator,' *IEEE Transactions on Energy Conversion*, Vol. 14, No. 3, pp. 394–401, September 1999.

[6] E.G. Marra and J.A. Pomilio, 'Induction-generator-based system providing regulated voltage with constant frequency,' *IEEE Transactions on Industrial Electronics*, Vol. 47, No. 4, pp. 908–914, August 2000.

[7] Z.M. Salameh and L.F. Kazda, 'Analysis of the steady state performance of the double output induction generator,' *IEEE Transactions on Energy Conversion*, Vol. EC-1, No. 1, pp. 26–32, March 1986.

[8] B.T. Ooi and R.A. David, 'Induction-generator/synchronous-condenser system for wind-turbine power,' *IEE Proceedings*, Vol. 126, No. 1, pp. 69–74, January 1979.

[9] K.A. Nigim, 'Static exciter for wound rotor induction machine,' *Conference Record of 16th IAS Annual Meeting*, Vol. 2, pp. 934–937, 1990.

[10] F. Giraud and Z.M. Salameh, 'Wind-driven, variable-speed, variable-frequency, double-output, induction generator,' *Electric Machines and Power Systems*, Vol. 26, No. 3, pp. 287–297, April 1998.

[11] L. Ouazene and G. McPherson, Jr, 'Analysis of the isolated induction generator,' *IEEE Transactions on Power Apparatus and Systems*, Vol. PAS-102, No. 8, pp. 2793–2798, August 1983.

[12] T.F. Chan, 'Analysis of self-excited induction generators using an iterative method,' *IEEE Transactions on Energy Conversion*, Vol. 10, No. 3, pp. 502–507, September 1995.

[13] B.D. Bunday, *Basic Optimisation Methods*, Edward Arnold, London, 1984.

[14] G.K. Dubey, *Power Semiconductor Controlled Drives*, Prentice Hall, Englewood Cliffs, NJ, 1989.

[15] K.J. Astrom and T. Hagglund, *PID Controllers: Theory, Design and Tuning*, Instrument Society of America, Research Triangle Park, NC, 1995.

7

PMSGs for Autonomous Power Systems

7.1 Introduction

A PMSG that possesses an inherent voltage compensation capability is desirable for maintaining a constant voltage for autonomous power system applications. In this chapter, it is demonstrated that the inverse saliency feature of a synchronous generator with an inset PM rotor can be exploited to improve the voltage regulation. Performance of the generator is computed based on the two-axis theory, and the conditions for achieving zero voltage regulation are deduced for the case of unity-power-factor loads as well as lagging-power-factor loads. The open-circuit voltage, direct-axis synchronous reactance and quadrature-axis synchronous reactance required in the analysis are accurately determined from the finite element method (FEM). A coupled circuit and field method will also be used for computing the generator load characteristic directly. The computed and experimental performance of a small prototype PMSG will be presented to validate the theoretical analysis.

7.2 Principle and Construction of PMSG with Inset Rotor

In an AC generator with a surface magnet rotor, the direct-axis (d-axis) synchronous reactance X_d and the quadrature-axis (q-axis) synchronous reactance X_q are approximately equal. Figure 7.1(a) shows the phasor diagram for unity-power-factor operation of a synchronous generator with a surface magnet rotor, where E is the open-circuit voltage, V is the terminal voltage, I is the armature current, R is the armature resistance and X_s ($= X_d = X_q$) is the synchronous reactance. It is

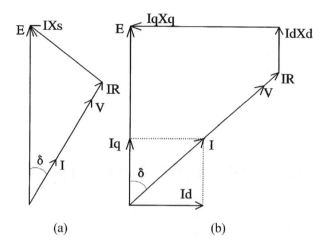

(a)   (b)

Figure 7.1   *Phasor diagram of PMSG when supplying a unity-power-factor load: (a) surface magnet rotor type; (b) generator with inverse saliency. Reproduced by permission of T. F. Chan, L. L. Lai and Lie-Tong Yan, 'Performance of a three-phase A.C. generator with inset NdFeB permanent-magnet rotor',* Transactions on Energy Conversion, **19**, *2004: 88–94. © (2004) IEEE*

obvious that V is less than E when the generator is on load. Figure 7.1(b) shows the phasor diagram of a synchronous generator that exhibits inverse saliency, i.e. a machine in which X_q is larger than X_d. It should be noted that the d-axis current I_d and q-axis current I_q, and hence the d-axis synchronous reactance drop $I_d X_d$ and q-axis synchronous reactance drop $I_q X_q$, depend on the load impedance. In general, a larger $I_q X_q$ drop causes the terminal voltage phasor V to fall further behind the open-circuit voltage phasor E. Under certain favourable conditions, the magnitude of V may be equal to, or even greater than, the magnitude of E. With a suitable ratio of X_q to X_d, it is possible to achieve zero voltage regulation at a particular load.

Figure 7.2 shows the cross-sectional view of a prototype PMSG with an inset rotor, the technical details of which are given in Appendix D.3. Each PM is arch shaped and is surface mounted on the rotor yoke using a suitable bonding material, while soft-iron pole pieces occupy the interpolar regions. The edges of each rotor magnet are tapered in such a way that its inner pole arc is wider than the outer pole arc. This design results in a better air gap flux density distribution and hence a more sinusoidal output voltage waveform. The gap between each magnet and the adjacent soft-iron pole piece is filled with epoxy for better mechanical strength. The rotor magnets are made of neodymium–iron–boron (Nd–Fe–B), a high-energy PM material whose recoil permeability is very close to that of air. This property results in a suppression of the d-axis flux linkage and hence X_q is larger than X_d, i.e. the generator exhibits inverse saliency. The desired X_q/X_d ratio can be obtained by an appropriate choice of the interpolar air gap length, the pole arcs of the permanent magnets and the widths of the interpolar soft-iron pole pieces.

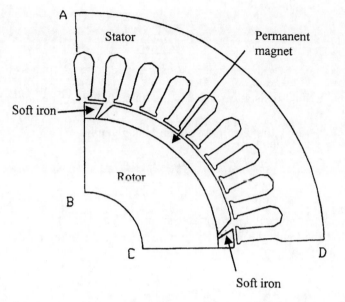

Figure 7.2 *Construction of a four-pole prototype PMSG with inset rotor, a quarter cross-section being shown. Reproduced by permission of T. F. Chan, L. L. Lai and Lie-Tong Yan, 'Performance of a three-phase A.C. generator with inset NdFeB permanent-magnet rotor',* Transactions on Energy Conversion, **19**, 2004: 88–94. © (2004) IEEE

7.3   Analysis for Unity-Power-Factor Loads

7.3.1   Analysis Using the Two-Axis Model

Heating and lighting loads usually predominate in autonomous power systems. In this section, this mode of operation is analysed for the prototype PMSG based on the two-axis model [1, 2]. Besides the prediction of the load characteristics, it is of interest to investigate the conditions at which the generator terminal voltage on load is equal to the open-circuit voltage, i.e. the generator operates with zero voltage regulation. Two cases will be considered: analysis with armature resistance neglected and analysis with armature resistance taken into consideration.

Armature resistance neglected. The armature resistance may be neglected when the machine rating is large. Referring to Figure 7.1(b), the following equations can be written:

$$V \cos \delta = E - I_d X_d \tag{7.1}$$

$$V \sin \delta = I_q X_q \tag{7.2}$$

$$I_d = I \sin \delta \tag{7.3}$$

$$I_q = I \cos \delta \tag{7.4}$$

$$I = \frac{V}{R_L} \tag{7.5}$$

$$I = \sqrt{I_d^2 + I_q^2} \tag{7.6}$$

where R_L is the load resistance and δ is the load angle (i.e. the angle between E and V).

Solving (7.1)–(7.6), the terminal voltage V is given by

$$V = \frac{E.R_L\sqrt{R_L^2 + X_q^2}}{R_L^2 + X_d X_q}. \tag{7.7}$$

The load characteristic can be computed from (7.5) and (7.7).

From (7.7), the load resistance R_L at which zero voltage regulation occurs (i.e. $V = E$) is

$$R_L = \frac{X_q}{\sqrt{r(r-2)}} \tag{7.8}$$

where r, the inverse saliency ratio, equals X_q/X_d.

Equation (7.8) indicates that when the inverse saliency ratio r exceeds 2, there exists a value of R_L that gives zero voltage regulation.

It can also be shown that the load angle δ at which zero voltage regulation occurs is given by

$$\tan\left(\frac{\delta}{2}\right) = \sqrt{\frac{r-2}{r}}. \tag{7.9}$$

Armature resistance considered. When the armature resistance cannot be neglected, a similar analysis can be carried out with reference to Figure 7.1(b). It can be shown that the load angle is given by

$$\tan\delta = \frac{I X_q}{V + IR} \tag{7.10}$$

while the terminal voltage V and the open-circuit voltage E are related by the following equation:

$$E = \frac{(V + IR)^2 + I^2 X_d X_q}{\sqrt{(V + IR)^2 + (IX_q)^2}}. \tag{7.11}$$

The load angle at which zero voltage regulation occurs is now given by

$$\tan\left(\frac{\delta}{2}\right) = \frac{R + X_d \tan\delta}{X_q - R \tan\delta}. \tag{7.12}$$

Equation (7.12) may be solved numerically to give the load angle δ. This approach, however, requires suitable initial estimates of δ to be selected in order to start the numerical procedure.

If the parameter $k = \tan(\delta /2)$ is introduced, (7.12) can be expanded to give the following cubic equation:

$$X_q k^3 + R k^2 + (2X_d - X_q)k + R = 0. \tag{7.13}$$

Standard closed forms of solution for (7.13) are available, e.g. by using Cardan's method [3]. Only positive real values of k give feasible operating points. Once k (and hence δ) is known, the load current at which zero voltage regulation occurs can be determined from (7.10).

As an example, consider a PMSG with the following parameters at rated speed [1]: $E = 51.7\,\text{V}$, $X_d = 14\,\Omega$, $X_q = 55\,\Omega$ and $R = 3.5\,\Omega$. Solution of (7.12) with these numerical values substituted yields the following positive real roots: $k_1 = 0.1374$ and $k_2 = 0.5875$. For root k_1, the load angle and armature current are $15.6°$e and 0.268A, respectively. For root k_2, the corresponding values are $60.9°$e and 1.9 A, respectively. There are thus two different loads at which zero voltage regulation is obtained. In practice, however, operation with the larger load would give more output power and a higher efficiency.

The solutions of (7.12) can also be determined graphically by the points of intersection between the functions f_1 and f_2, where $f_1 = \tan(\delta /2)$ and $f_2 = (R + X_d \tan\delta)/(X_q - R \tan\delta)$, as illustrated in Figure 7.3 for the PMSG in [1]. When $R = 3.5\,\Omega$, f_1 and f_2 intersect at two points. It is apparent that the function f_2

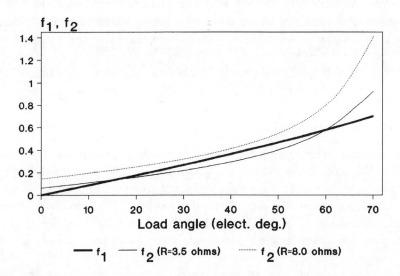

Figure 7.3 *Graphical solution of (7.12) for determining the load angle δ. Reproduced by permission of T. F. Chan, L. L. Lai and Lie-Tong Yan, 'Performance of a three-phase A.C. generator with inset NdFeB permanent-magnet rotor',* Transactions on Energy Conversion, **19**, *2004: 88–94.* © *(2004) IEEE*

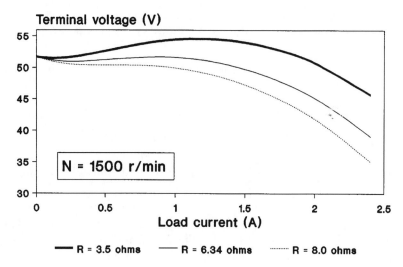

Figure 7.4 *Effect of armature resistance R on the load characteristics of PMSG with E = 51.7 V, X_d = 14 Ω, X_q = 55 Ω. Reproduced by permission of T. F. Chan, L. L. Lai and Lie-Tong Yan, 'Performance of a three-phase A.C. generator with inset NdFeB permanent-magnet rotor', Transactions on Energy Conversion,* **19**, *2004: 88–94. © (2004) IEEE*

is displaced upwards with an increase in R, causing the separation between the intersection points to decrease. When $R = 8\,\Omega$, f_2 is larger than f_1 for all values of δ. There is no point of intersection and hence zero voltage regulation cannot be achieved. At some critical value of R, the two roots are equal, implying that zero voltage regulation is obtained at only one value of load current. For the given machine, the critical value of R was found to be 6.34 Ω using a simple search method.

Figure 7.4 shows the load characteristics of the generator for different values of R.

7.3.2   Design Considerations

For the electrical machine designer, it is of interest to determine, for a given speed, armature resistance R and d-axis synchronous reactance X_d, the value of q-axis synchronous reactance X_q (or r) such that zero voltage regulation occurs at a specific armature current. To simplify the derivation, (7.11) is rewritten in per-unit form, the no-load voltage E being taken as unity. Thus, with $V = E = 1.0\,\text{p.u.}$, the following equation is obtained:

$$\frac{\alpha^2 + I_{pu}^2 X_d X_q}{\sqrt{\alpha^2 + (I_{pu} X_q)^2}} = 1 \qquad (7.14)$$

where I_{pu} is the per-unit armature current and $\alpha = 1 + I_{pu} R$.

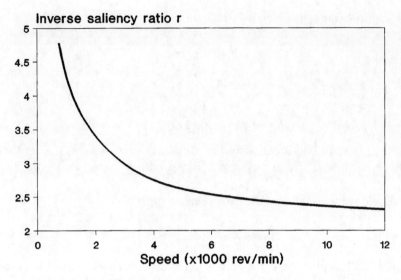

Figure 7.5 *Inverse saliency ratio r to give zero voltage regulation in PMSG at different speeds. Reproduced by permission of T. F. Chan, L. L. Lai and Lie-Tong Yan, 'Performance of a three-phase A.C. generator with inset NdFeB permanent-magnet rotor',* Transactions on Energy Conversion, **19**, *2004: 88–94.* © *(2004) IEEE*

From (7.14), the inverse saliency ratio r is given by

$$r = \frac{\alpha}{\beta} \cdot \frac{\alpha . I_{pu} X_d + \sqrt{\alpha^2 - \beta}}{I_{pu} X_d} \tag{7.15}$$

where $\beta = 1 - (I_{pu} X_d)^2$.

Figure 7.5 shows the variation of r with speed to yield zero voltage regulation at rated current for a PMSG with $R = 0.295 \, \Omega$ and $X_d = 0.88 \, \Omega$ (at the nominal frequency of 50 Hz). It is apparent that for low-speed operation, the generator has to be designed with a large inverse saliency ratio in order to achieve zero voltage regulation.

For a PMSG with negligible armature resistance, (7.15) is reduced to the following:

$$r = \frac{2}{\beta} = \frac{2}{1 - (I_{pu} X_d)^2}. \tag{7.16}$$

Since the per-unit value of X_d is independent of speed, the value of r to give zero voltage regulation is the same for all speeds.

It is also of interest to study the effect of speed on the armature resistance to yield zero voltage regulation. Solving (7.14) for α, one obtains

$$\alpha^2 = \frac{1}{2} \left[-(2r I_{pu}^2 X_d^2 - 1) \pm \sqrt{1 + 4r^2 I_{pu}^2 X_d^2 - 4r I_{pu}^2 X_d^2} \right]. \tag{7.17}$$

The per-unit resistance is given by

$$R = \frac{\alpha - 1}{I_{pu}}. \tag{7.18}$$

But

$$R = \frac{R_{actual}}{Z_{base,n} \cdot b} \tag{7.19}$$

where R_{actual} is the ohmic value of the armature resistance, $Z_{base,n}$ is the base impedance at nominal frequency and b is the ratio of actual speed to the base speed.

The ohmic value of the armature resistance is thus

$$R_{actual} = \frac{\alpha - 1}{I_{pu}} \cdot Z_{base,n} \cdot b. \tag{7.20}$$

Equation (7.20) implies that, for zero voltage regulation at a specific current, a larger armature resistance can be tolerated when the generator operates at a higher speed.

7.3.3   Computed Results

For comparison purposes the load current characteristics of the prototype PMSG with inset rotor and a surface magnet PMSG with similar design were computed using the two-axis model. The machine parameters required were determined from the FEM (Section 7.5) and are listed there in Table 7.1. For easy comparison, the voltages have been normalized to the corresponding no-load voltages. Figure 7.6 shows that at the nominal speed of 1500 r/min, the full-load voltage drops in the

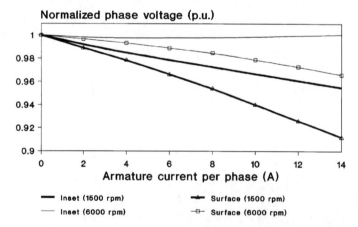

Figure 7.6   *Computed load characteristics of synchronous generators with inset and surface-mounted PM rotors. Reproduced by permission of T. F. Chan, L. L. Lai and Lie-Tong Yan, 'Performance of a three-phase A.C. generator with inset NdFeB permanent-magnet rotor', Transactions on Energy Conversion, **19**, 2004: 88–94. © (2004) IEEE*

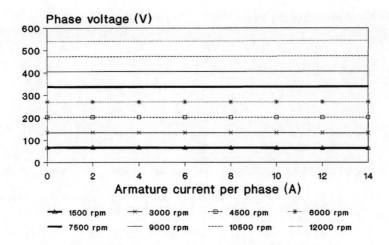

Figure 7.7 *Computed load characteristics of prototype PMSG with inset rotor. Reproduced by permission of T. F. Chan, L. L. Lai and Lie-Tong Yan, 'Performance of a three-phase A.C. generator with inset NdFeB permanent-magnet rotor',* Transactions on Energy Conversion, **19**, *2004: 88–94.* © *(2004) IEEE*

generators with inset PM rotor and surface-mounted PM rotor are 4.4 % and 8.4 % respectively. At a speed of 6000 r/min, the generator with inset PM rotor exhibits a nearly level voltage characteristic, with zero voltage drop occurring at full load. The computed results have confirmed that the inset PM rotor construction is effective in improving the voltage regulation.

Figure 7.7 shows the computed voltage–current characteristics of the prototype generator when operating at different speeds. It is observed that the voltage compensation due to inverse saliency increases with speed, which is consistent with the derivation in Section 7.3.2. Above 6000 r/min, the terminal voltage rises with increase of load, i.e. negative voltage regulation can be obtained.

7.3.4 Experimental Results

Figure 7.8 shows the experimental performance characteristics of the prototype PMSG when driven at the nominal speed (1500 r/min) and supplying a unity-power-factor load. The base voltage and current per phase are taken as 63.5 V and 13.3 A respectively. At the rated current, the voltage drop of the generator is 5.9 % and the power output is 1.0 p.u. (2.5 kW) at an efficiency of 86.2 %.

Figure 7.9 shows the experimental and computed load characteristics of the PMSG when supplying a balanced unity-power-factor load and driven at different speeds. The characteristics are nearly level and parallel to each other, hence the machine is suitable for use as a constant voltage generator. The close agreement between the computed and experimental results in Figure 7.9 confirms the accuracy of the parameters obtained from the FEM.

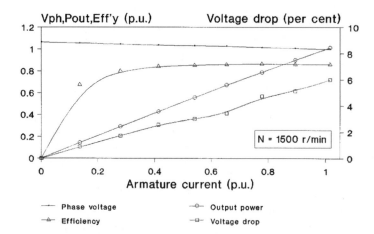

Figure 7.8   *Experimental performance characteristics of prototype PMSG at rated speed. Reproduced by permission of T. F. Chan, L. L. Lai and Lie-Tong Yan, 'Performance of a three-phase A.C. generator with inset NdFeB permanent-magnet rotor', Transactions on Energy Conversion, **19**, 2004: 88–94. © (2004) IEEE*

7.3.5   Summary

The analysis and performance of a three-phase synchronous generator with inset PM rotor have been presented in this section. Particular emphasis has been placed on the conditions for achieving zero voltage regulation when the generator is supplying an isolated resistive load. It is demonstrated that the voltage regulation is significantly improved as a result of the inverse saliency feature of the inset PM

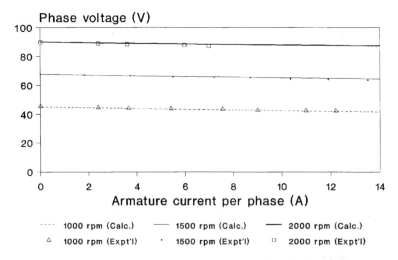

Figure 7.9   *Load characteristics of prototype PMSG at different speeds. Reproduced by permission of T. F. Chan, L. L. Lai and Lie-Tong Yan, 'Performance of a three-phase A.C. generator with inset NdFeB permanent-magnet rotor', Transactions on Energy Conversion, **19**, 2004: 88–94. © (2004) IEEE*

rotor construction. Experiments performed on a 2.5 kVA prototype generator have confirmed the accuracy of the theoretical analysis.

7.4 A Comprehensive Analysis

In this section, a comprehensive analysis of the prototype PMSG with inset rotor will be presented. The analysis, based on the two-axis model, will be extended to include the general case of the lagging-power-factor load. The conditions for achieving zero voltage regulation, extremum points in the load characteristic and maximum power output will be deduced analytically and a saturated two-axis model will also be proposed.

7.4.1 Basic Equations and Analysis

Figure 7.10 shows the phasor diagram of the PMSG when it is supplying a lagging-power-factor load.

It is convenient to use the load impedance Z_L as a variable in the performance analysis for isolated operation. With reference to Figure 7.10, the following equations may be written for generator operation with a lagging-power-factor load:

$$V \cos \delta = E - I_d X_d - I_q R \qquad (7.21)$$

$$V \sin \delta = I_q X_q - I_d R \qquad (7.22)$$

$$I_d = I \sin(\delta + \phi) \qquad (7.23)$$

$$I_q = I \cos(\delta + \phi) \qquad (7.24)$$

$$V = I.Z_L \qquad (7.25)$$

where Z_L is the load impedance and ϕ is the power factor load angle.

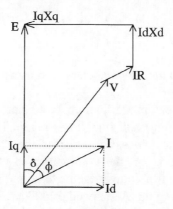

Figure 7.10 *Phasor diagram of PMSG when supplying a lagging-power-factor load. Reproduced by permission of T.-F. Chan and L. L. Lai, 'Permanent-magnet synchronous generator with inset rotor for autonomous power system applications',* Proceedings – Generation, Transmission & Distribution, **151**, 2004: 597–603

Substituting (7.23), (7.24) and (7.25) into (7.22),

$$Z_L \sin \delta = X_q \cos(\delta + \phi) - R \sin(\delta + \phi). \tag{7.26}$$

Expanding the right hand side of (7.26) and rearranging terms, the following equation may be deduced:

$$\tan \delta = \frac{X_q \cos \phi - R \sin \phi}{Z_L + X_q \sin \phi + R \cos \phi}. \tag{7.27}$$

For a given load impedance Z_L and load impedance angle ϕ, the load angle δ can be computed from (7.27). Substitution of the value of δ into (7.21) gives the terminal voltage V as follows:

$$V = \frac{E \cdot Z_L}{Z_L \cos \delta + R \cos(\delta + \phi) + X_d \sin(\delta + \phi)}. \tag{7.28}$$

The load current (or armature current) I is given by

$$I = \frac{E}{Z_L \cos \delta + R \cos(\delta + \phi) + X_d \sin(\delta + \phi)}. \tag{7.29}$$

Equations (7.28) and (7.29) enable the load characteristics of the PMSG to be evaluated.

It is also possible to express the terminal voltage V directly in terms of the load impedance Z_L. Eliminating δ from (7.27) and (7.28), the following equation is obtained:

$$V = E \cdot \frac{Z_L \sqrt{Z_L^2 + g_1. Z_L + g_2}}{Z_L^2 + g_3. Z_L + g_4} \tag{7.30}$$

where

$$g_1 = 2(X_q \sin \phi + R\cos\phi), \quad g_2 = R^2 + X_q^2, \quad g_3 = 2R \cos \phi + (X_d + X_q) \sin \phi,$$

$$g_4 = R^2 + X_d X_q.$$

The load characteristic can therefore be computed using (7.25) and (7.30) without having to determine δ first.

Figure 7.11 shows the computed load characteristics of the prototype PMSG at nominal speed (1500 r/min) and four times nominal speed (6000 r/min). It is apparent that the voltage compensation effect of the PMSG is weaker when the power factor load is lagging. At the nominal speed (1500 r/min), the full-load voltage drop is 4.7% when the power factor load is unity and 15.7% when the power factor load is 0.8 lagging. At a speed of 6000 r/min, the generator exhibits a nearly level load characteristic when the power factor load is unity, with zero voltage drop occurring at full load. At 0.8 power factor lagging, the corresponding voltage drop is 11%.

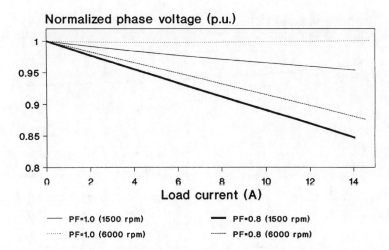

Figure 7.11 *Computed load characteristics of PMSG with the following machine parameters at nominal speed: $E = 66.44$ V, $R = 0.295\,\Omega$, $X_d = 0.88\,\Omega$, $X_q = 2.23\,\Omega$. Reproduced by permission of T.-F. Chan and L. L. Lai, 'Permanent-magnet synchronous generator with inset rotor for autonomous power system applications',* Proceedings – Generation, Transmission & Distribution, **151**, 2004: 597–603. © (2004) IET

Figure 7.12 shows the effect of inverse saliency ratio r on the load characteristics of the PMSG when supplying a unity-power-factor load. For this investigation, it is assumed that E, R and X_d of the generator are the same as the corresponding values of the prototype generator. It is observed that the optimum value of r is 4, at which a

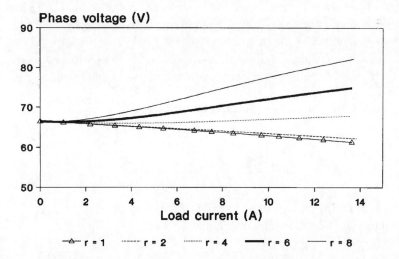

Figure 7.12 *Effect of inverse saliency ratio r on the load characteristics of a PMSG operating at nominal speed and supplying a unity-power-factor load: $E = 66.44$ V, $R = 0.295\,\Omega$, $X_d = 0.88\,\Omega$. Reproduced by permission of T.-F. Chan and L. L. Lai, 'Permanent-magnet synchronous generator with inset rotor for autonomous power system applications',* Proceedings – Generation, Transmission & Distribution, **151**, 2004: 597–603. © (2004) IET

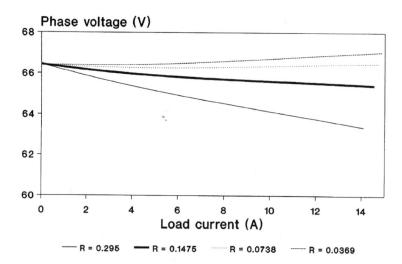

Figure 7.13   *Effect of armature resistance R on the load characteristics of a PMSG operat-*
ing at nominal speed and supplying a unity-power-factor load: E = 66.44 V, X_d = 0.88 Ω,
X_q = 2.23 Ω. Reproduced by permission of T.-F. Chan and L. L. Lai, 'Permanent-magnet syn-
chronous generator with inset rotor for autonomous power system applications', Proceedings
– Generation, Transmission & Distribution, **151,** *2004: 597–603.* © *(2004) IET*

nearly level load characteristic is obtained. For values of r exceeding 4, the terminal
voltage increases with load current over most of the practical current range.

Figure 7.13 shows the effect of armature resistance R on the load characteristics
of the prototype PMSG when supplying a unity-power-factor load. A reduction in
R results in a smaller voltage drop. A comparison between Figure 7.12 and Figure
7.13 shows that reducing R to a quarter of the original value has approximately the
same effect as increasing r by four times. When R is smaller than one-quarter of
the original value, negative voltage regulation can also be obtained.

7.4.2   Conditions for Zero Voltage Regulation

From Figure 7.10, it can be shown that, for $V = E$, the load angle δ is given by

$$\tan\left(\frac{\delta}{2}\right) = \frac{R + X_d \tan(\delta + \phi)}{X_q - R \tan(\delta + \phi)}. \tag{7.31}$$

If we let $k = \tan(\delta/2)$, $r = X_q/X_d$, $m_1 = R/X_d$ and $m_2 = \tan\phi$, (7.31) can be
expanded to give the following cubic equation in k:

$$a_3 k^3 + a_2 k^2 + a_1 k + a_0 = 0 \tag{7.32}$$

where

$$a_0 = m_1 + m_2, \quad a_1 = 2 - r - m_1 m_2, \quad a_2 = m_1 + m_2(2r - 1),$$
$$a_3 = r - m_1 m_2.$$

Equation (7.32) is similar in form to (7.13) and hence may also be solved using Cardan's method [3]. Positive real roots of (7.32), if they exist, indicate that zero voltage regulation may be achieved. The corresponding value of δ and load current can subsequently be computed.

It is also possible to deduce the zero voltage regulation condition with the load impedance Z_L as the variable. Putting $V = E$, (7.30) may be expanded to give the following cubic polynomial in Z_L:

$$b_3.Z_L^3 + b_2.Z_L^2 + b_1.Z_L + b_0 = 0 \qquad (7.33)$$

where

$$b_0 = g_4^2, \quad b_1 = 2g_3.g_4, \quad b_2 = g_3^2 + 2g_4 - g_2, \quad b_3 = 2g_3 - g_1.$$

As an illustration, (7.33) was solved for the prototype generator supplying a unity-power-factor load at 6000 r/min. Under this condition, $E = 265.6$ V, $R = 0.295\,\Omega$, $X_d = 3.52\,\Omega$, $X_q = 8.92\,\Omega$ and $\cos\phi = 1.0$. Solution of (7.33) gave two positive roots, namely 20.74 and 13.1. When $Z_L = 20.74\,\Omega$, $\delta = 22.98°$e and $I = 12.8$ A; when $Z_L = 13.1\,\Omega$, $\delta = 33.62°$e and $I = 20.24$ A.

Figure 7.14 shows the effect of rotor speed on the load angle that gives zero voltage regulation and the corresponding load current I when the prototype generator is supplying a unity-power-factor load. When the rotor speed is higher than the critical value of 5687 r/min, there are two values of δ (or I) at which zero voltage regulation occurs. When the rotor speed is less than the critical value, no positive real roots of (7.32) or (7.33) exist and hence a zero voltage regulation condition

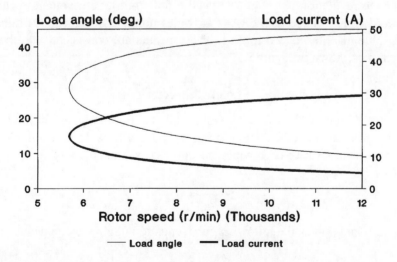

Figure 7.14 *Effect of rotor speed on the load angle that gives zero voltage regulation when the prototype generator is supplying a unity-power-factor load and the corresponding load current. Reproduced by permission of T.-F. Chan and L. L. Lai, 'Permanent-magnet synchronous generator with inset rotor for autonomous power system applications', Proceedings – Generation, Transmission & Distribution, **151**, 2004: 597–603. © (2004) IET*

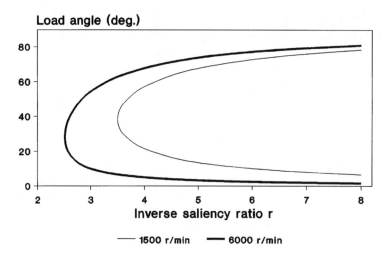

Figure 7.15   *Effect of inverse saliency ratio r on the load angle that gives zero voltage regulation at nominal speed for a generator with $R = 0.295\,\Omega$, $X_d = 0.88\,\Omega$. Reproduced by permission of T.-F. Chan and L. L. Lai, 'Permanent-magnet synchronous generator with inset rotor for autonomous power system applications', Proceedings – Generation, Transmission & Distribution, **151**, 2004: 597–603. © (2004) IET*

does not exist. It should be noted that at the critical speed, (7.32) or (7.33) gives repeated positive roots.

Figure 7.15 shows the effect of inverse saliency ratio on the load angle that gives zero voltage regulation. At nominal speed (1500 r/min), zero voltage regulation is achieved when r exceeds 3.5. At four times nominal speed (6000 r/min), the critical value of r is reduced to 2.51. These results are consistent with the load characteristics shown in Figure 7.12.

7.4.3   Extremum Points in the Load Characteristic

The results in Figures 7.11–7.13 also suggest that extremum points might exist in the load characteristic of the PMSG. If Z_L is used as a variable, this condition will occur when the derivative of V with respect of Z_L is equal to zero, i.e.

$$\frac{dV}{dZ_L} = 0. \tag{7.34}$$

From (7.30) and (7.34), the following cubic equation in Z_L may be deduced:

$$c_3.Z_L^3 + c_2.Z_L^2 + c_1.Z_L + c_0 = 0 \tag{7.35}$$

where

$$c_0 = 2g_2g_4, \quad c_1 = 3g_1g_4, \quad c_2 = g_1g_3 + 4g_4 - 2g_2, \quad c_3 = 2g_3 - g_1.$$

Solution of (7.35) for the prototype generator operating at a speed of 6000 r/min and supplying a unity-power-factor load also gives two positive real roots, namely. 50.8 and 15.8. When $Z_L = 50.8\,\Omega$, $\delta = 9.9°$e, $I = 5.2$ A and $V = 264.9$ V; when $Z_L = 15.8\,\Omega$, $\delta = 29.0°$e, $I = 16.9$ A and $V = 265.9$ V.

The results in Sections 7.4.2 and 7.4.3 suggest that the load characteristic of the prototype PMSG at 6000 r/min and supplying a unity-power-factor load comprises an initial concave-upward section and a subsequent concave-downward section. At light loads the armature resistance causes a drop in terminal voltage, but as the load current increases, the voltage compensation due to inverse saliency becomes dominant and it gradually offsets the armature resistance drop. This explains the presence of a 'saddle' in the load characteristic.

7.4.4 Power–Load Angle Relationship

Since the terminal voltage V varies with load, the power–load angle characteristic of a PMSG under isolated operation is different from that under infinite busbar operation. For a three-phase generator supplying an isolated load, the total output power is given by

$$P = \frac{3V^2}{Z_L}.\cos\phi\,. \tag{7.36}$$

Substituting (7.26) and (7.28) into (7.36), the power–load angle equation for the PMSG can be deduced:

$$P = \frac{3E^2\cos\phi[X_q\cos(\delta+\phi) - R\sin(\delta+\phi)]\sin\delta}{[X_q\cos\delta.\cos(\delta+\phi) - R\sin\phi + X_d\sin\delta.\sin(\delta+\phi)]^2}. \tag{7.37}$$

Figure 7.16 shows the effect of inverse saliency ratio r on the power–load angle characteristics of a PMSG with $E = 66.44$ V, $R = 0.295\,\Omega$, $X_d = 0.88\,\Omega$ at nominal speed (1500 r/min) and supplying a unity-power-factor load. For a given value of r, a maximum power condition occurs at a load angle which is less than $90°$e. The maximum power output from the generator increases with increase in r, but at light loads the value of δ for a given output power is larger.

The maximum power condition cannot easily be deduced from (7.37). To simplify the mathematical derivation, the power is first expressed in terms of Z_L as follows:

$$P = 3E^2\cos\phi.\frac{Z_L(Z_L^2 + g_1.Z_L + g_2)}{(Z_L^2 + g_3.Z_L + g_4)^2}. \tag{7.38}$$

For a given power factor angle ϕ, maximum power output occurs when

$$\frac{dP}{dZ_L} = 0. \tag{7.39}$$

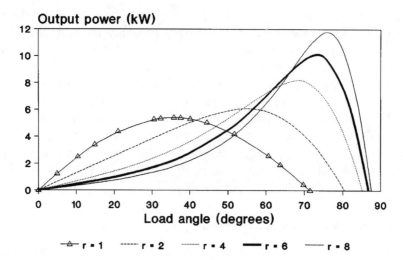

Figure 7.16 *Effect of inverse saliency ratio r on the power–load angle characteristics of a PMSG operating at nominal speed and supplying a unity-power-factor load: E = 66.44 V, R = 0.295 Ω, X_d = 0.88 Ω. Reproduced by permission of T.-F. Chan and L. L. Lai, 'Permanent-magnet synchronous generator with inset rotor for autonomous power system applications',* Proceedings – Generation, Transmission & Distribution, **151**, 2004: 597–603. © (2004) IET

From (7.38) and (7.39), the following quartic polynomial equation is obtained:

$$d_4.Z_L^4 + d_3.Z_L^3 + d_2.Z_L^2 + d_1.Z_L + d_0 = 0 \qquad (7.40)$$

where

$$d_0 = -g_2 g_4, \quad d = g_2 g_3 - 2g_1 g_4, \quad d_2 = 3(g_2 - g_4), \quad d_3 = 2g_1 - g_3,$$
$$d_4 = 1.$$

Equation (7.40) may be solved in a straightforward manner using Ferrari's method [3]. Again only positive real roots yield feasible extremum points.

7.4.5   The Saturated Two-Axis Model

A saturated two-axis model for the PMSG with inset rotor is now proposed based on the results obtained from a 2-D FEM (to be elaborated in Section 7.5). Figure 7.24 below shows the computed variations X_d and X_q as a function of the exciting currents I_d and I_q. For representation in a computer program, the variations may be approximated by the following describing equations:

$$X_d = \begin{cases} 0.86 + 0.003 I_d, & I_d < 10 \\ 0.89, & I_d \geq 10 \end{cases} \qquad (7.41)$$

$$X_q = \begin{cases} 2.23, & I_q < 6 \\ 2.605 - 0.0625 I_q, & I_q \geq 6. \end{cases} \qquad (7.42)$$

The analysis using the saturated two-axis model can be summarized as follows:

1. Specify initial values of X_d and X_q. For convenience, the unsaturated values could be used.
2. For a given load impedance Z_L, power factor load angle ϕ and rotor speed, evaluate the terminal voltage V using the method presented in Section 7.4.1.
3. Compute the d-axis current I_d and q-axis current I_q, using (7.23) and (7.24).
4. Compute the new values of X_d and X_q from (7.41) and (7.42), respectively.
5. Repeat steps 2 to 4 until the values of V in successive iterations are less than a specified small value, say 0.000 001.
6. Compute the load voltage and generator performance using the final value of terminal voltage and the corresponding load angle.

Figure 7.17 shows the computed and experimental load characteristics of the prototype PMSG when driven at the nominal speed (1500 r/min). At heavy loads, the voltage compensation effect due to inverse saliency is partly offset by saturation in the q-axis magnetic circuit. For unity-power-factor loads, the saturation effect is noticeable for load currents exceeding 6 A. It is observed that the analysis based on the saturated two-axis model gives a more accurate prediction of the load characteristic compared with the two-axis model with fixed parameters, especially under heavy load conditions. For loads at 0.8 power factor lagging, however, both methods give practically the same results. This is due to the fact that I_q is smaller and hence the effect of q-axis saturation is less pronounced.

The good correlation between the computed and experimental results confirms the validity of the analysis based on the saturated two-axis model.

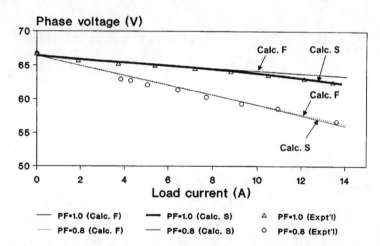

Figure 7.17 *Load characteristics of prototype PMSG at nominal speed (Calc. F, calculated using fixed values of X_d and X_q; Calc. S, calculated using saturated two-axis model). Reproduced by permission of T.-F. Chan and L. L. Lai, 'Permanent-magnet synchronous generator with inset rotor for autonomous power system applications', Proceedings – Generation, Transmission & Distribution,* **151***, 2004: 597–603.* © *(2004) IET*

7.4.6 Summary

This section has presented a comprehensive analysis of a three-phase PMSG with inset rotor. Relevant equations for lagging-power-factor loads are developed based on the two-axis model. The conditions for achieving zero voltage regulation, extremum points on the load characteristic and maximum power output have been deduced analytically. It is shown that the saturated two-axis model gives a more accurate prediction of the load characteristic at heavy loads.

7.5 Computation of Synchronous Reactances

7.5.1 Analysis Based on FEM

The synchronous reactances X_d and X_q may be computed from an analysis based on a 2-D FEM. To compute the magnetic field distribution, eight-node quadrilateral elements [4] are used in the mesh formation and FEM computations in order to reduce the number of elements required while maintaining a reasonably high computational accuracy. Advantage is taken of the symmetry in the machine configuration so that the solution region need only cover one-quarter of the cross-section of the prototype generator (Figure 7.2). The FEM mesh used consists of 108 elements and 429 nodes.

The rotor PM is modelled by an equivalent current density J_c that also accounts for the curvature effects of the magnets. Magnetic vector potential A is taken as the variable in the FEM formulation.

Magnetic hysteresis in the iron cores is neglected, i.e. the nominal magnetization curve is used for modelling magnetic saturation. Figure 7.18 shows the magnetization curve for the ferromagnetic material of the prototype generator.

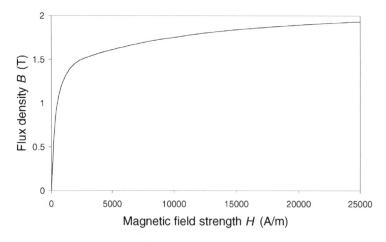

Figure 7.18   *Magnetization curve of prototype PMSG*

With reference to Figure 7.2, the basic field equation and boundary conditions of the PMSG may be written as follows [4]:

$$\frac{\partial}{\partial x}\left(\frac{1}{\mu}\frac{\partial A}{\partial x}\right) + \frac{\partial}{\partial y}\left(\frac{1}{\mu}\frac{\partial A}{\partial y}\right) = -J \tag{7.43}$$

$$\frac{1}{\mu_1}\frac{\partial A}{\partial n}\bigg|_{L^-} - \frac{1}{\mu_2}\frac{\partial A}{\partial n}\bigg|_{L^+} = J_c \tag{7.44}$$

$$A|_{AD} = A|_{BC} \tag{7.45}$$

$$A|_{AB} = -A|_{DC} \tag{7.46}$$

where $ABCD$ = boundary surface defined in Figure 7.2;
       J = externally impressed current density;
       J_c = equivalent surface current density of PM material;
       L = surface of PM along the direction of the magnetization;
       L^+ = edge adjacent to L and outside the PM material;
       L^- = edge adjacent to L and inside the PM material;
       μ = permeability;
       μ_1 = permeability of PM material;
       μ_2 = permeability of air.

Depending on the region being considered, (7.43) may need to be modified slightly. In sourceless regions, as in the case of the stator core and the rotor yoke, J should be equated to zero.

The energy functional is

$$E(A) = \iint_{\Omega}\left(\int_0^B \frac{B}{\mu}\cdot dB - A\cdot J\right)dx\cdot dy - \int_L J_c\cdot A.dl' \tag{7.47}$$

where B is the flux density and dl' is an infinitesimal segment along edge L.

After discretization and functional minimization, the following matrix equation is obtained:

$$[K][A] = [R] \tag{7.48}$$

where $[K] = n \times n$ coefficient (stiffness) matrix;
       $[A] = n \times 1$ column matrix of nodal magnetic vector potentials;
       $[R] = n \times 1$ right-hand-side column matrix containing known terms;
       n = number of nodes.

After solving (7.48) to give A, the flux density at any point (x, y) may be computed as follows:

$$B = \sqrt{B_x^2 + B_y^2} \tag{7.49}$$

$$\begin{bmatrix} B_x \\ B_y \end{bmatrix} = [J]^{-1} \begin{bmatrix} \dfrac{\partial N_1}{\partial \eta} & \cdots & \dfrac{\partial N_8}{\partial \eta} \\[2mm] \dfrac{\partial N_1}{\partial \xi} & \cdots & \dfrac{\partial N_8}{\partial \xi} \end{bmatrix} \begin{bmatrix} A_1 \\ \cdot \\ \cdot \\ \cdot \\ A_8 \end{bmatrix} \tag{7.50}$$

where $[J]^{-1}$ = inverse of Jacobian matrix;
$N_1, \ldots, N_8$ = fundamental functions;
ξ, η = local coordinates.

7.5.2 Computation of X_d and X_q

For computing X_d or X_q, the usual definition of air gap magnetizing reactance is used, but the effect of rotor magnets on the flux produced by the exciting current in each axis has to be accounted for.

To compute X_d, the d-axis of the rotor is aligned with the resultant magnetomotive force (m.m.f.) axis of the armature winding and a specified balanced three-phase current is allowed to flow. With the surface current density J_c set to that corresponding to the remanent flux density, the magnetic field distribution is computed and the values of permeability in the elements are determined. Using these values of magnetic permeability, the magnetic field distribution is recalculated with J_c set to zero and the resultant flux linkage is evaluated. X_d is then determined from the ratio of the induced e.m.f. to the armature current, plus the leakage reactance of the end winding X_e. The value of X_d computed in this manner has thus included the effect of the rotor magnets.

The flux linkage in the ith coil in the d-axis winding is

$$\Psi_i = w.\phi_i \tag{7.51}$$

where

$$\phi_i = \int_{S_i} \overline{B}.d\overline{s} = \int \Delta \times \overline{A}.d\overline{s} = \int_l \overline{A}.d\overline{l} = (A_i - A_i')l_{fe}. \tag{7.52}$$

The induced e.m.f. in the d-axis is

$$E_d = 4.44f \left(\sum_{i=1}^{pg} \psi_i \right) \Big/ c. \tag{7.53}$$

X_d is then computed as the ratio of the induced e.m.f. in the winding to the corresponding current, i.e.

$$X_d = \frac{E_d}{I_d} + X_e \qquad (7.54)$$

where A_i, A_i' = equivalent vector potential at the coil sides of the ith coil;
   c = number of parallel paths;
   B = flux density;
   E_d = induced e.m.f. in the d-axis winding;
   f − frequency of induced e.m.f.;
   g = number of slots per pole per phase;
   I_d = d-axis armature current;
   l = boundary of plane S_i;
   l_{fe} = effective length of the iron core;
   p = number of poles;
   S_i = area formed by the coil sides of the ith coil;
   w = turns per coil;
   X_e = leakage reactance of the end winding;
   ψ_i = flux linkage in the ith coil;
   ϕ_i = flux in the ith coil.

X_q can be computed in a similar manner, but the q-axis of the rotor should be aligned with the resultant m.m.f. axis of the armature winding.

7.5.3   Computed Results

The above algorithms were implemented in a FORTRAN program with reference to the prototype generator shown in Figure 7.2. Computations were also performed on an identical generator with a surface magnet rotor (i.e. one without the interpolar soft-iron pole pieces). The per-phase machine parameters, computed for an exciting current of 6 A, are listed in Table 7.1.

Figure 7.19 shows the flux plot of the prototype PMSG on no load. Since there is no armature current, all the flux is contributed by the rotor magnets. It is observed that the presence of soft-iron pole pieces in the interpolar axis results in an additional

Table 7.1   *Per-phase parameters of prototype generator X_d and X_q computed at a current of 6 A*

Rotor type	X_d (Ω)	X_q (Ω)	E (V)
Inset PM rotor	0.88	2.23	66.44
Surface magnet rotor	0.73	0.74	67.67

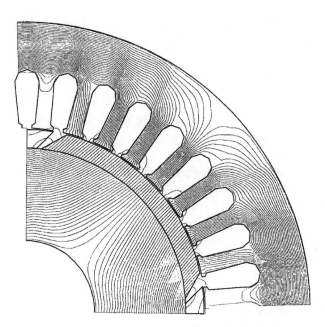

Figure 7.19   *Flux density distribution due to rotor PM alone. Reproduced by permission of T.-F. Chan and L. L. Lai, 'Permanent-magnet synchronous generator with inset rotor for autonomous power system applications'*, Proceedings – Generation, Transmission & Distribution, **151**, *2004: 597–603.* © *(2004) IET*

leakage rotor flux, hence the no-load voltage is slightly less than that of an equivalent generator with surface magnet rotor.

Figure 7.20 shows the flux density distribution due to the d-axis current alone. Most of the d-axis flux traverses a large effective air gap, either through the radial length of the magnet or through the stator slots, hence the d-axis flux linkage is naturally suppressed. On the other hand, the q-axis flux plot in Figure 7.21 shows that a large portion of the q-axis flux traverses the low-reluctance path provided by the soft-iron pole pieces. The origin of inverse saliency in the inset rotor is thus clearly demonstrated. In Figure 7.21, regions of high flux density can be identified, implying that local magnetic saturation is prominent in the q-axis.

Figure 7.22 and Figure 7.23 show respectively the composite flux plot (i.e. flux plot with combined rotor and stator excitation) of the PMSG when excited with d-axis and q-axis current. It is observed that the q-axis current gives rise to a more irregular flux density distribution and the flux density distribution is no-longer symmetrical as in the case of d-axis excitation.

By repeating the FEM computations presented in Section 7.5.2 over the practical range of stator exciting current, the variation of X_d with I_d, as well as the variation of X_q with I_q, can be obtained.

The computed values of X_d and X_q for the prototype generator are shown, respectively, by the triangular and circular symbols in Figure 7.24. It is seen that the inset rotor construction results in a remarkable increase in X_q. The inverse saliency

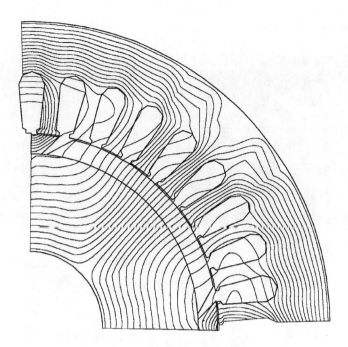

Figure 7.20 *Flux density distribution due to d-axis current alone. Reproduced by permission of T.-F. Chan and L. L. Lai, 'Permanent-magnet synchronous generator with inset rotor for autonomous power system applications'*, Proceedings – Generation, Transmission & Distribution, **151**, 2004: 597–603. © (2004) IET

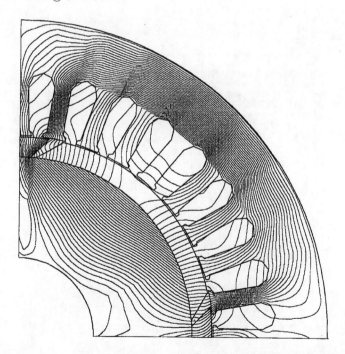

Figure 7.21 *Flux density distribution due to q-axis current alone. Reproduced by permission of T.-F. Chan and L. L. Lai, 'Permanent-magnet synchronous generator with inset rotor for autonomous power system applications'*, Proceedings – Generation, Transmission & Distribution, **151**, 2004: 597–603. © (2004) IET

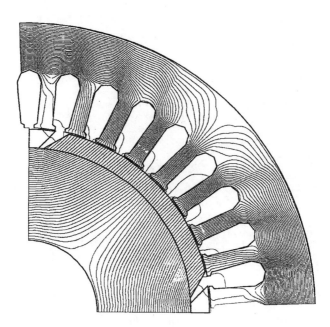

Figure 7.22 *Composite flux plot of prototype generator when excited by d-axis armature current. Reproduced by permission of T. F. Chan, L. L. Lai and Lie-Tong Yan, 'Performance of a three-phase A.C. generator with inset NdFeB permanent-magnet rotor',* Transactions on Energy Conversion, **19**, 2004: 88–94. © (2004) IEEE

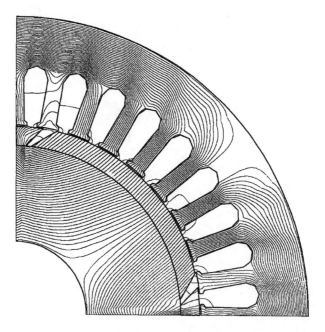

Figure 7.23 *Composite flux plot of prototype generator when excited by q-axis current. Reproduced by permission of T. F. Chan, L. L. Lai and Lie-Tong Yan, 'Performance of a three-phase A.C. generator with inset NdFeB permanent-magnet rotor',* Transactions on Energy Conversion, **19**, 2004: 88–94. © (2004) IEEE

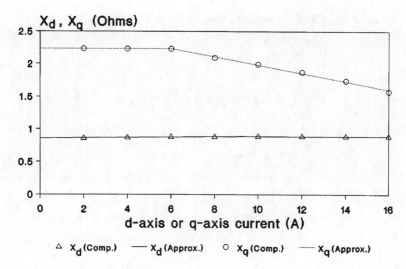

Figure 7.24   *Variations of X_d and X_q with stator exciting current computed from FEM, and the corresponding piecewise-linearized approximations. Reproduced by permission of T.-F. Chan and L. L. Lai, 'Permanent-magnet synchronous generator with inset rotor for autonomous power system applications', Proceedings – Generation, Transmission & Distribution, 151, 2004: 597–603.* © *(2004) IET*

ratio r of the generator is approximately equal to 2.53 under unsaturated conditions. Due to magnetic saturation in the q-axis, X_q decreases as I_q increases. X_d, on the other hand, increases slightly with I_d due to the demagnetizing effect of I_d on the rotor magnet. The computed results indicate that the saturation characteristic of the PMSG with inset rotor is different from that of an interior-type PM generator [1].

7.5.4   Summary

A method for computing the d-axis and q-axis synchronous reactances for the PMSG has been presented. Magnetic saturation, including the effect of rotor magnets, has been included in the analysis. The origin of inverse saliency is clearly demonstrated from the flux plots. The results are useful for performance evaluation of the PMSG, especially in the development of a saturated two-axis model.

7.6   Analysis using Time-Stepping 2-D FEM

7.6.1   Machine Model and Assumptions

In previous sections, the PMSG with inset rotor has been analysed using the two-axis model, the values of X_d and X_q required being determined from a field analysis based on the FEM. This approach is computationally efficient since only one-quarter of the machine cross-section (for the four-pole prototype generator) needs to be modelled and the flux density distribution need only be computed for a single

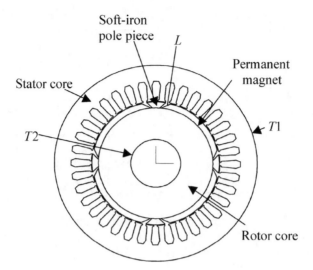

Figure 7.25 *Cross-sectional view of prototype PMSG with inset rotor construction. Repro-duced by permission of T.-F. Chan, L. L. Lai and L.-T. Yan, 'Analysis of a stand-alone permanent-magnet synchronous generator using a time-stepping coupled field-circut method', Proceed-ings – Electric Power Applications,* **152***, 2005: 1459–1467.* © *(2005) IET*

position of the rotor. It should be pointed out, however, that the two-axis model is essentially an approximate method and mixed magnetic saturation conditions that exist in the actual machine cannot be accounted for.

In Chapter 4, a time-stepping 2-D FEM has been applied for the analysis of a grid-connected IG with the Steinmetz connection. In this section, this approach will be pursued again for the performance analysis of a PMSG with an inset rotor.

The solution region has to be extended to the complete cross-section of the PM generator since the axisymmetric boundary condition is no longer valid. As shown in Figure 7.25, the field solution region Ω is bounded by the outer periphery $T1$ of the stator core and the inner periphery $T2$ of the rotor yoke (which coincides with the outer surface of the rotor shaft). It is assumed that all the flux is confined within the region Ω.

As in Chapter 4, first-order triangular elements and linear interpolation functions are used, but only balanced operation at constant speed will be considered.

7.6.2　Coupled Circuit and Field Analysis

The field equations given in Section 7.5 are also applicable for the present anal-ysis, except for the boundaries with zero magnetic potential. With reference to Figure 7.25, the boundary condition is given by

$$A|_{T1} = A|_{T2} = 0. \tag{7.55}$$

The matrix equation (7.48) in the field domain has to be coupled to the external circuit. For a PMSG supplying an isolated load, the external circuit comprises the armature resistance R, the armature end-winding leakage inductance L_e, the

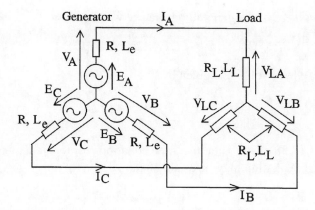

Figure 7.26 *PMSG supplying an isolated, balanced, three-wire star-connected load. Reproduced by permission of T.-F. Chan, L. L. Lai and L.-T. Yan, 'Analysis of a stand-alone permanent-magnet synchronous generator using a time-stepping coupled field-circuit method', Proceedings – Electric Power Applications,* **152**, *2005: 1459–1467.* © *(2005) IET*

load resistance R_L and the load inductance L_L. Figure 7.26 shows the circuit connection of a star-connected PMSG which supplies a balanced star-connected load. Six circuit variables are introduced, namely the resultant generated e.m.f.s E_A, E_B and E_C and the phase currents I_A, I_B and I_C. It should be noted that E_A, E_B and E_C can each be expressed in terms of the nodal magnetic potentials, hence it remains to establish the equations for the phase currents.

For the mesh containing A-phase and B-phase in Figure 7.26,

$$E_A - E_B = R_t I_A + L_t \frac{dI_A}{dt} - R_t I_B - L_t \frac{dI_B}{dt} \tag{7.56}$$

where $R_t = R + R_L$ and $L_t = L_e + L_L$.

Rewriting (7.56) in differential form,

$$E_A - E_B = R_t I_A + L_t \frac{(I_A - I_A')}{\Delta t} - R_t I_B - L_t \frac{(I_B - I_B')}{\Delta t} \tag{7.57}$$

where I_A' = value of A-phase current in the previous time step;
I_B' = value of B-phase current in the previous time step.

Eqn. (7.57) can further be expressed in the following form:

$$m_1 I_A - m_1 I_B - E_A + E_B = m_2 I_A' - m_2 I_B' \tag{7.58}$$

where $m_1 = R_t + L_t/\Delta t$ and $m_2 = L_t/\Delta t$.

In a similar manner, for the mesh containing B-phase and C-phase,

$$m_1 I_B - m_1 I_C - E_B + E_C = m_2 I_B' - m_2 I_C'. \tag{7.59}$$

For a three-wire system, the line currents must satisfy the following constraint:

$$I_A + I_B + I_C = 0. \tag{7.60}$$

The A-phase voltage of the generator armature winding is

$$V_A = E_A - RI_A - m_2(I_A - I'_A) \tag{7.61}$$

and the A-phase load voltage is given by

$$V_{LA} = R_L I_A + m_3(I_A - I'_A) \tag{7.62}$$

where $m_3 = L_L/\Delta t$.

From (7.48), (7.58)–(7.60) and the three equations for the generated e.m.f.s, the following matrix equation may be established:

$$[K'][A'] = [R'] \tag{7.63}$$

where $[K'] = n' \times n'$ modified stiffness matrix;
         $[A'] = n' \times 1$ column matrix containing all the field and circuit variables;
         $[R'] = n' \times 1$ column matrix containing known terms;
         $n' = n + 6$.

Solution of (7.63) enables the magnetic vector potentials A, flux density, stator e.m.f.s, stator currents and the terminal voltage to be determined. It should be

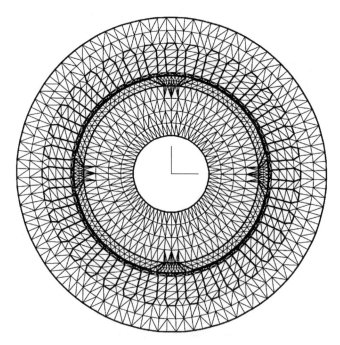

Figure 7.27 *Mesh formation for 2-D FEM analysis of PMSG with inset rotor. Reproduced by permission of T.-F. Chan, L. L. Lai and L.-T. Yan, 'Analysis of a stand-alone permanent-magnet synchronous generator using a time-stepping coupled field-circut method', Proceedings – Electric Power Applications, **152**, 2005: 1459–1467. © (2005) IET*

noted that the synchronous reactances need not be evaluated throughout the solution procedure.

Figure 7.27 shows the mesh formation for the proposed analysis using the time-stepping 2-D FEM. There are 2816 nodes and 5384 elements. The number of nodes with zero magnetic vector potential (i.e. nodes that fall on boundaries $T1$ and $T2$) total 248, hence the total number of field variables is $2816 - 248 = 2568$. Since there are six additional circuit variables, the total number of variables in the coupled circuit and field formulation is $2568 + 6 = 2574$.

7.6.3 Magnetic Saturation Consideration

In order to account for the effect of saturation, a magnetic nonlinearity subprogram has been incorporated in the FEM solver. For this purpose, the magnetization curve shown in Figure 7.18 is used. Figure 7.28 shows the flowchart of the proposed time-stepping 2-D FEM coupled with the external circuit equations. With the magnetic permeability μ of each element initialized to the appropriate value (for regions containing ferromagnetic material, the unsaturated value of μ is used), the stiffness matrix $[K]$ is formed and modified according to the boundary conditions and the external circuit configuration. For each time step, the modified system of equations is solved and the flux densities in all the elements are evaluated from the nodal vector potentials. Using the magnetization curve of the magnetic material, the corresponding values of permeability μ (and hence the reluctivity υ) are updated. The computations are repeated until the iteration error $\Delta\xi^2$ or the reluctivity error $\Delta\upsilon^2$ is within the specified values, or the specified number of iterations has been reached. The iteration error $\Delta\xi^2$ is defined as follows:

$$\Delta\xi^2 = \sum_{i=1}^{n'}\left[A_i'^{(k+1)} - A_i'^{(k)}\right]^2 \tag{7.64}$$

where $A_i'^{(k+1)}$ = value of ith variable of the modified column matrix $[A']$ in the $(k + 1)$th iteration;

$A_i'^{(k)}$ = value of ith variable of the modified column matrix $[A']$ in the kth iteration.

The reluctivity error $\Delta\upsilon^2$ is defined as

$$\Delta\upsilon^2 = \sum_{i=1}^{N_e}\left[\upsilon_{i,new}^{(k)} - \upsilon_i^{(k)}\right]^2 \tag{7.65}$$

where $\upsilon_i^{(k)}$ = value of reluctivity of the ith element at the beginning of the kth iteration;

$\upsilon_{i,new}^{(k)}$ = value of reluctivity at the end of the kth iteration;

N_e = number of elements.

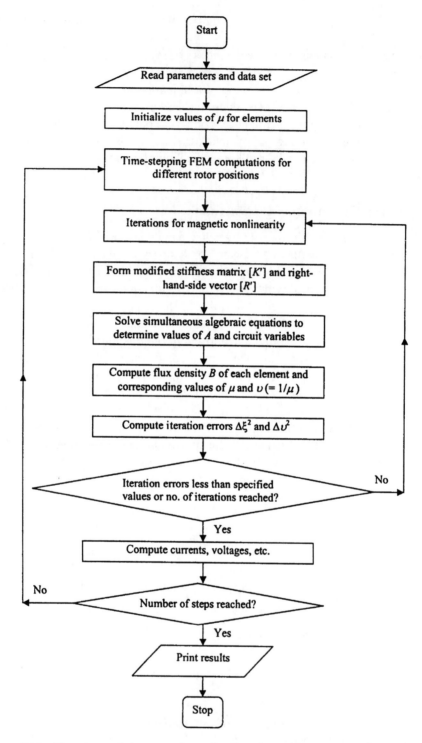

Figure 7.28 Flowchart of time-stepping 2-D FEM for performance analysis of PMSG considering magnetic saturation. Reproduced by permission of T.-F. Chan, L. L. Lai and L.-T. Yan, 'Analysis of a stand-alone permanent-magnet synchronous generator using a time-stepping coupled field-circut method', Proceedings – Electric Power Applications, **152**, 2005: 1459–1467. © (2005) IET

Arelaxation factor may be applied for updating the values of υ for the next iteration:

$$\upsilon_i^{(k+1)} = \upsilon_i^{(k)} + W\left(\upsilon_{i,new}^{(k)} - \upsilon_i^{(k)}\right) \tag{7.66}$$

where $\upsilon_i^{(k+1)}$ = value of reluctivity at the beginning of the $(k+1)$th itercation;
   W = relaxation factor.

7.6.4 Computed Results

The above algorithms were implemented in a FORTRAN program that incorporates the FEM computations, preprocessing and postprocessing routines. In the solver, each time step was equivalent to one mechanical degree. For each load impedance or operating point, the FEM computations were performed over 1440 steps (i.e. four complete revolutions of the rotor) to allow the system quantities to reach steady state. For the magnetic nonlinearity subprogram, the maximum number of iterations was specified to be five. This choice was a compromise between the desired computational accuracy and the computation time required for a complete solution. An acceleration factor of 0.2 was found to give satisfactory convergence. The solution procedure was time consuming due to the large number of variables and also the large number of steps involved. On a Pentium 4 personal computer with a clock speed of 1.6 GHz, the computation time for one operating point was approximately 30 h.

Figure 7.29 shows the computed flux plot of the PMSG under no-load conditions. Since there is no armature current, all the flux is contributed by the rotor magnets.

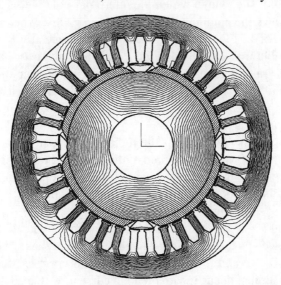

Figure 7.29   *Flux plot of PMSG with inset rotor under no-load condition. Reproduced by permission of T.-F. Chan, L. L. Lai and L.-T. Yan, 'Analysis of a stand-alone permanent-magnet synchronous generator using a time-stepping coupled field-circut method',* Proceedings – Electric Power Applications, *152, 2005: 1459–1467.* © *(2005) IET*

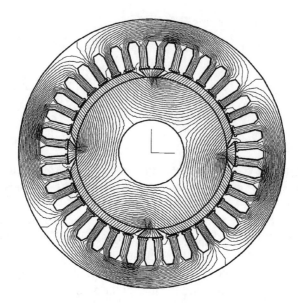

Figure 7.30   *Flux plot of PMSG with inset rotor when supplying a resistive load of 4.71 Ω per phase. Reproduced by permission of T.-F. Chan, L. L. Lai and L.-T. Yan, 'Analysis of a stand-alone permanent-magnet synchronous generator using a time-stepping coupled field-circuit method', Proceedings – Electric Power Applications, **152**, 2005: 1459–1467. © (2005) IET*

A small rotor leakage flux can be observed to flow across the soft-iron pole pieces, hence the no-load voltage is slightly less than that of an equivalent generator with surface magnet rotor (i.e. one in which the interpolar pole pieces are removed).

Figure 7.30 shows the computed flux plot of the PMSG when it is supplying a pure resistive load of 4.71 Ω per phase. Due to the presence of soft-iron pole pieces, there is considerable magnetic flux in the q-axis and magnetic saturation can be significant in the q-axis region, e.g. the base of each soft-iron pole piece and the stator teeth adjacent to it. This condition cannot easily be analysed by using the two-axis model and hence the proposed time-stepping 2-D FEM that includes the effect of magnetic saturation is justified.

Figure 7.31 shows the variation of the radial component B_n of the no-load air gap flux density in the PMSG, computed at the mean radius of the air gap. Due to the stator slot openings, undulations are produced in the B_n waveform. It is also observed that B_n is practically zero in the q-axis region, which agrees with the computed flux plot shown in Figure 7.29.

Figure 7.32 and Figure 7.33 show the variations in B_n when the PMSG is supplying a unity-power-factor load of 9.1 Ω and 4.71 Ω per phase (which correspond to load currents of 7.1 A and 13.2 A), respectively. It is observed that the q-axis flux density increases almost linearly with the load current, with a consequent increase in the effective flux per pole. Since the flux in the q-axis depends upon the armature current, the flux density compensation effect increases with the load current. For a well-designed generator, it is possible for the voltage compensation partially to

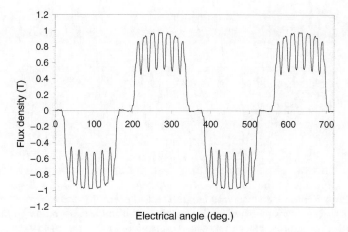

Figure 7.31 *Computed variation of B_n when the PMSG is on no load. Reproduced by permission of T.-F. Chan, L. L. Lai and L.-T. Yan, 'Analysis of a stand-alone permanent-magnet synchronous generator using a time-stepping coupled field-circuit method', Proceedings – Electric Power Applications, **152**, 2005: 1459–1467.* © *(2005) IET*

offset the voltage drop due to the armature leakage impedance and d-axis armature reaction reactance. A reduced voltage regulation, or even zero voltage regulation, can therefore be achieved.

Figure 7.34 shows the computed phase and line voltage waveforms of the PMSG on no load, while Figure 7.35 shows the computed line voltage and line current when the PMSG is supplying full-load current at unity power factor. Since a resistive

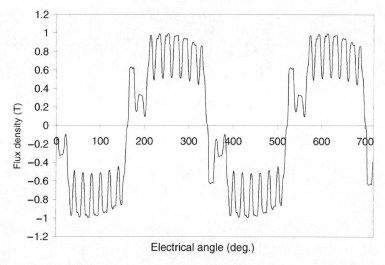

Figure 7.32 *Computed variation of B_n when the PMSG is supplying a load resistance of 9.1 Ω per phase. Reproduced by permission of T.-F. Chan, L. L. Lai and L.-T. Yan, 'Analysis of a stand-alone permanent-magnet synchronous generator using a time-stepping coupled field-circut method', Proceedings – Electric Power Applications, **152**, 2005: 1459–1467.* © *(2005) IET*

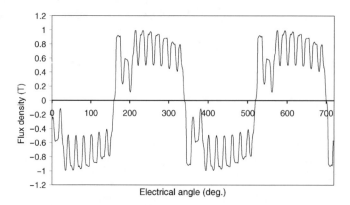

Figure 7.33 *Computed variation of B_n when the PMSG is supplying a load resistance of 4.71 Ω per phase. Reproduced by permission of T.-F. Chan, L. L. Lai and L.-T. Yan, 'Analysis of a stand-alone permanent-magnet synchronous generator using a time-stepping coupled field-circut method',* Proceedings – Electric Power Applications, *152, 2005: 1459–1467. © (2005) IET*

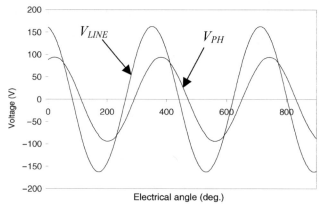

Figure 7.34 *Computed line voltage (V_{LINE}) and phase voltage (V_{PH}) of the PMSG at no load.*

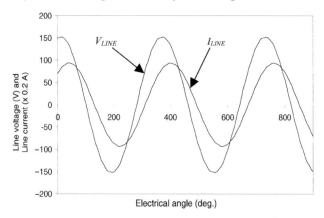

Figure 7.35 *Computed line voltage (V_{LINE}) and line current (I_{LINE}) of the PMSG when supplying full-load current at unity power factor. Reproduced by permission of T.-F. Chan, L. L. Lai and L.-T. Yan, 'Analysis of a stand-alone permanent-magnet synchronous generator using a time-stepping coupled field-circut method',* Proceedings – Electric Power Applications, *152, 2005: 1459–1467. © (2005) IET*

load is being supplied, the phase voltage V_{PH} is in phase with the phase (line) current I_{LINE} and hence the line voltage V_{LINE} leads I_{LINE} by $30°$e. Despite the use of the inset rotor construction, very small harmonic distortion is observed from the waveforms. It is seen that the slot harmonics have been suppressed to a large extent as a result of the distribution of armature winding and the use of short-pitched coils.

7.6.5 Experimental Verification

A load test was performed on the prototype PMSG in order to validate the results computed using the coupled circuit time-stepping 2-D FEM. Figure 7.36 shows the computed and experimental load characteristics obtained. The computed characteristic is nearly level when the PMSG is supplying a unity-power-factor load and the full-load voltage drop is only 6.4%. When the power factor load is 0.8 lagging, the voltage drop at full load increases to 12.7%. The voltage compensation capability of the PMSG with inset rotor is therefore better for unity-power-factor loads. Very good agreement between the computed and experimental characteristics has been obtained, confirming the accuracy of the proposed coupled circuit and field method.

Figure 7.37 shows the no-load line voltage (V_{LINE}) and phase voltage (V_{PH}) waveforms of the prototype generator at rated speed. Figure 7.38 shows the line voltage (V_{LINE}) and line current (I_{LINE}) of the prototype generator at full load. The experimental waveforms are very similar to the computed waveforms shown in Figure 7.34 and Figure 7.35. It is interesting to note that there is less harmonic distortion

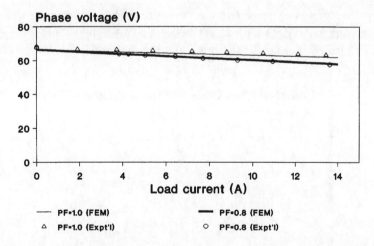

Figure 7.36 *Computed and experimental load characteristics of PMSG when supplying unity-power-factor and 0.8 lagging-power-factor loads at nominal speed. Reproduced by permission of T.-F. Chan, L. L. Lai and L.-T. Yan, 'Analysis of a stand-alone permanent-magnet synchronous generator using a time-stepping coupled field-circut method', Proceedings – Electric Power Applications, 152, 2005: 1459–1467.* © *(2005) IET*

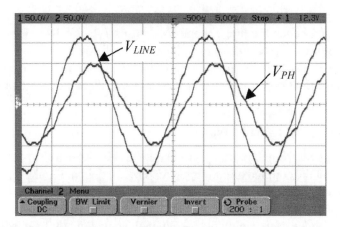

Figure 7.37   *Waveforms of the phase voltage (V_{PH}) and line voltage (V_{LINE}) of the PMSG on no load (voltage scale: 50 V/div; time scale: 5 ms/div)*

in the experimental waveforms at full load, and the load current is practically sinusoidal. A harmonic analysis of the waveforms in Figure 7.38 revealed that the total harmonic distortion (THD) in the line voltage was only 2.1 %. The fifth and the seventh harmonics in the full-load line voltage waveform were dominant and were equal to 0.9 % and 1.5 %, respectively. The harmonic content of the PMSG is thus acceptable for practical applications.

7.6.6   Summary

A coupled circuit and field approach has been attempted for the performance analysis of a three-phase PMSG with inset rotor and a two-dimensional finite element method (2-D FEM) is used for the field solution. The solver developed has also

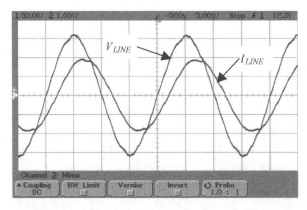

Figure 7.38   *Line voltage (V_{LINE}) and line current (I_{LINE}) waveforms of PMSG at full load. (voltage scale: 50 V/div; current scale: 10 A /div; time scale: 5 ms/div)*

accounted for the effect of saturation on the air gap flux density and the load characteristics. The experimental voltage and current waveforms are similar to those computed from the 2-D FEM, while good agreement between the computed and experimental load characteristics has been obtained. These observations confirm the validity of the FEM model and the accuracy of the solution procedure.

References

[1] B.J. Chalmers, 'Performance of interior type permanent-magnet alternator,' *IEE Proceedings–Electric Power Applications*, Vol. 141, No. 4, pp. 186–190, July 1994.
[2] M.A. Rahman, A.M. Osheiba, T.S. Radwan and E.S. Abdin, 'Modelling and controller design of an isolated diesel engine PM synchronous generator,' *IEEE Transactions on Energy Conversion*, Vol. 11, No. 2, pp. 324–330, June 1996.
[3] M.V. Sweet, *Algebra, Geometry and Trigonometry in Science, Engineering and Mathematics*, Ellis Horwood, Chichester, 1984.
[4] Pi-Zhang Chen, Lie-Tong Yan and Ruo-Ping Yao, *Theory and Computation of Electromagnetic Fields in Electrical Machines*, Science Press, China, 1986.

8

Conclusions

8.1 Accomplishments of the Book

This book has explored the feasibility of using the induction generator (IG) and the permanent magnet synchronous generator for distributed and autonomous power system applications.

The study of the IG has focused on single-phase operation of a three-phase machine. For the grid-connected IG, a systematic analysis based on the method of symmetrical components enables the steady-state performance of various phase balancing schemes to be investigated. The analysis is subsequently extended to a three-phase IG with the Smith connection (SMIG) and Smith's Mode C circuit. The feasibility of microcontroller-based multi-mode operation of the SMIG is also studied.

A coupled circuit and field approach for analysing a single-phase grid-connected IG with the Steinmetz connection is proposed. The book has presented the general methodology for coupling the asymmetrical stator circuit equations to the field equations based on a time-stepping 2-D FEM. A detailed rotor circuit model is also introduced in order to account for the non-uniform current distribution in the rotor winding. The results computed from the FEM and the symmetrical components method are compared with the experimental data.

Another major accomplishment of the book is the development and analysis of practical SEIG schemes for asymmetrically connected and single-phase loads. By using the method of symmetrical components and an effective search algorithm, the steady-state performance of various SEIG schemes that employ three-phase machines can be analysed. For the special case of single-phase loads, the modified Steinmetz connection (MSC) and simplified Steinmetz connection (SSC) have been proposed and investigated. For improving the voltage regulation, an SRSEIG scheme that possesses automatic voltage-regulating as well as good phase balancing capability is developed. By an appropriate choice of the shunt and series capacitances, a nearly level load characteristic of the SRSEIG may be obtained.

The capacitor sizing problem of a self-excited induction generator (SEIG) is also addressed with reference to the Steinmetz connection. The results throw light on the proper selection of excitation capacitance for securing self-excitation in other SEIG schemes.

The SMSEIG, namely an SEIG with the Smith connection, is proposed and investigated quantitatively for the first time. This scheme is particularly suitable for supplying single-phase unity-power-factor loads. It is demonstrated that the voltage unbalance factor of the SMSEIG is not sensitive to change in rotor speed. This suggests that, once the SMSEIG has been balanced at a particular speed, its performance over a wide speed range will remain satisfactory. The same argument may be extended to other SEIG schemes.

The use of a self-excited slip-ring induction generator (SESRIG) permits voltage and frequency control over a wide speed range and hence the excitation scheme is suitable for use with a variable speed wind energy conversion system. The voltage and frequency control characteristics are deduced and practical implementation of a closed-loop control scheme using a rotor resistance chopper circuit is described. With a properly tuned PI controller, satisfactory dynamic response of the SESRIG to speed and load changes has been obtained.

To address the recent developments in small-scale autonomous power systems, a prototype permanent magnet synchronous generator with inset rotor is studied. Due to its inverse saliency characteristic, the generator is capable of automatic voltage compensation. The analysis is first developed for the resistive load case but is subsequently extended to the general lagging-power-factor load case. Of particular interest is the deduction of the conditions for zero voltage regulation and the corresponding shape of the load characteristic. For the prototype generator, it is found that a nearly level load characteristic is obtained when it operates at around 6000 r/min, implying that the prototype generator is suitable for use in a direct-coupled, engine-driven power system. The parameters necessary for the two-axis model are obtained from a 2-D FEM and a saturated model is proposed in order to improve the computational accuracy. A coupled circuit and field approach that enables mixed d-axis and q-axis saturation conditions to be analysed is also proposed. The computed flux density distribution shows clearly the effect of load on the interpolar flux density and hence the resulting load characteristic.

Experimental work on the prototype PMSG has confirmed the accuracy of the two-axis model and the parameters computed from the FEM. Furthermore, the waveforms of voltages and currents are practically sinusoidal, a fact that renders the prototype PMSG a suitable machine for supplying isolated loads.

8.2 Future Work

The IG and SEIG schemes presented in the book are by no means exhaustive. The methodologies presented in Chapters 3 and 5 may readily be applied to other feasible phase balancing schemes. A phasor diagram analysis, for example, has revealed that perfect phase balance in the SMIG could be realized with purely resistive phase converters. The resulting system could be useful in applications where a large amount of local load has to be supplied. Variable speed operation of the SMSEIG may also be extended to other SEIG schemes, such as the SESRIG, with the turbine characteristic duly included in the analysis. The various phase balancing schemes for the IG and SEIG could also be applied to the reluctance generator (RG). A more refined analysis, however, needs to be developed, particularly the accurate modelling and determination of negative-sequence circuit parameters.

The coupled circuit and field analysis for the IG has been studied with reference to a grid-connected machine with the Steinmetz connection. With appropriate modification of the circuit equations and the use of a proper magnetization curve, the analysis could be extended to other grid-connected IG schemes or SEIG schemes. The results of field analysis could also be processed to yield the transient performance, such as load changes, switching of phase converters, etc.

The rotor resistance controller for the SESRIG scheme presented in Chapter 6 could be replaced by a slip energy recovery circuit that returns the slip energy to the stator side. Alternatively, the doubly fed induction generator (DFIG) [1], in which the rotor is fed by a PWM inverter, may be employed. Various vector control methods (both sensorless or with rotor position sensors) could be developed for voltage and frequency control.

The coupled circuit and field analysis developed for the PMSG could be extended to certain operational problems, such as nonlinear loads, load switching, terminal short circuits and unbalanced loads. New machine configurations for specific applications should also be studied, e.g. direct-coupled PMSGs for wind energy conversion systems.

With more and more renewable generation being introduced into the existing power networks, system integration aspects need to be studied. These may include, for example, new converter topologies, power factor correction, reduction in harmonic distortion, system stability, generator system dynamic interactions and plant economics.

Reference

[1] R.S. Pena, G.M. Asher and J.C. Clare, 'A doubly fed induction generator using back to back PWM converters supplying an isolated load from a variable speed wind turbine,' *IEE Proceedings – Electric Power and Applications*, Vol. 143, No. 5, pp. 380–387, September 1996.

Appendix A
Analysis for IG and SEIG

A.1 Symmetrical Components Equations for IG

The symmetrical components equations for phase voltages and phase currents are [1]

$$\begin{bmatrix} V_A \\ V_B \\ V_C \end{bmatrix} = \frac{1}{\sqrt{3}} \begin{bmatrix} 1 & 1 & 1 \\ 1 & h^2 & h \\ 1 & h & h^2 \end{bmatrix} \begin{bmatrix} V_0 \\ V_p \\ V_n \end{bmatrix} \tag{A.1}$$

$$\begin{bmatrix} I_A \\ I_B \\ I_C \end{bmatrix} = \frac{1}{\sqrt{3}} \begin{bmatrix} 1 & 1 & 1 \\ 1 & h^2 & h \\ 1 & h & h^2 \end{bmatrix} \begin{bmatrix} I_0 \\ I_p \\ I_n \end{bmatrix} \tag{A.2}$$

where h is the complex operator $e^{j2\pi/3}$.

It is assumed that induction machines with symmetrical three-phase stator windings are used. For these machines, Wagner and Evans [2] have established that currents of different sequences do not react upon each other. It follows that when voltages of a given sequence are applied to the induction machine, only currents of the same sequence are produced. The sequence voltages and currents are therefore decoupled, and the following relationships are valid:

$$I_0 = V_0/Z_0 = V_0 Y_0 \tag{A.3}$$

$$I_p = V_p/Z_p = V_p Y_p \tag{A.4}$$

$$I_n = V_n/Z_n = V_n Y_n. \tag{A.5}$$

In (A.3)–(A.5), Z_0, Z_p and Z_n are, respectively, the zero-, positive-, and

Distributed Generation: Induction and Permanent Magnet Generators L. L. Lai and T. F. Chan
© 2007 John Wiley & Sons, Ltd

negative-sequence impedance of the induction machine while Y_0, Y_p and Y_n are, respectively, the zero-, positive-, and negative-sequence admittance of the induction machine.

A.2   Positive-Sequence and Negative-Sequence Circuits of IG

Zero-sequence quantities are absent in the IG schemes investigated, hence only the positive-sequence and negative-sequence circuits need to be considered. Figure A.1 and Figure A.2 show, respectively, the positive-sequence and negative-sequence equivalent circuits of the grid-connected IG. For convenience, the *motor* convention has been adopted for the direction of current, i.e. the reference direction of current I_p or I_n is into the stator winding and the voltage V_p or V_n is considered to be applied to the machine. The advantage of this approach is that the symmetrical components analysis developed for induction motors could be applied directly. This convention will be used whenever the method of symmetrical components is used.

In Figure A.1 and Figure A.2, s_p and s_n denote, respectively, the per-unit slip of the rotor with respect to the positive-sequence rotating field and the negative-sequence rotating field. It is obvious that $s_n = 2 - s_p$.

From Figure A.1, the positive-sequence impedance Z_p and admittance Y_p are given by

$$Z_p = |Z_p| \angle \phi_p = (R_1 + jX_1) + R_c // jX_m // \left(\frac{R_p}{s_p} + jX_2 \right) \tag{A.6}$$

$$Y_p = \frac{1}{Z_p} = |Y_p| \angle -\phi_p \tag{A.7}$$

where $|Z_p|$ is the magnitude of Z_p, $|Y_p|$ is the magnitude of Y_p, and ϕ_p is the positive-sequence impedance angle. Both $|Z_p|$ and ϕ_p are functions of the per-unit slip s_p. For normal operation as an IG, s_p is a small negative number, hence ϕ_p will in general vary between $\pi/2$ and π rad. Accordingly, the positive-sequence current I_p lags the positive-sequence voltage V_p by an angle greater than $\pi/2$ rad. The *input* power is therefore negative, implying that the machine is delivering power to the stator circuit.

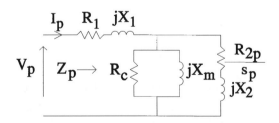

Figure A.1   *Positive-sequence equivalent circuit of grid-connected IG*

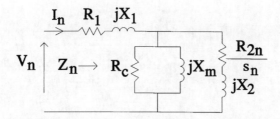

Figure A.2   *Negative-sequence equivalent circuit of grid-connected IG*

The negative-sequence impedance Z_n and admittance Y_n can likewise be computed from Figure A.2.

For isolated (or self-excited) operation, the frequency of generated e.m.f. of the IG may not be constant. To facilitate analysis, it is convenient to refer the circuit quantities to the base (rated) frequency f_{base} by introducing the following parameters [3]:

1. Per-unit frequency a, defined by

$$a = \text{(actual frequency)/(base frequency)} = f/f_{base}.$$

2. Per-unit speed b, defined by

$$b = \text{(actual rotor speed)/(synchronous speed corresponding to base frequency)}$$

$$= n_r/n_{base} = n_r/(f_{base}/p)$$

where p is the number of pole pairs and n_r is the rotor speed.

The positive-sequence and negative-sequence impedances of the IG for isolated operation are shown in Figure A.3 and Figure A.4.

A.3 V_p and V_n for IG with Dual-Phase Converters

From (3.2) and (A.1),

$$V_0 = 0 \text{ and } I_0 = 0 \tag{A.8}$$

i.e. zero-sequence quantities are absent from the system.

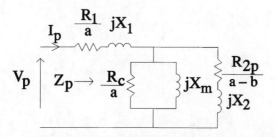

Figure A.3   *Positive-sequence equivalent circuit of IG for isolated operation*

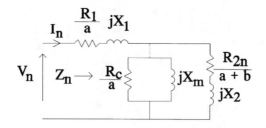

Figure A.4   *Negative-sequence equivalent circuit of IG for isolated operation*

Substituting (A.3)–(A.5) into (A.2), (3.3) and (3.4),

$$I_A = \frac{1}{\sqrt{3}}\left(I_p + I_n\right) = \frac{1}{\sqrt{3}}\left(V_p Y_p + V_n Y_n\right) \tag{A.9}$$

$$I_B = \frac{1}{\sqrt{3}}\left(h^2 I_p + h I_n\right) = \frac{1}{\sqrt{3}}\left(h^2 V_p Y_p + h V_n Y_n\right) \tag{A.10}$$

$$I_C = \frac{1}{\sqrt{3}}\left(h I_p + h^2 I_n\right) = \frac{1}{\sqrt{3}}\left(h V_p Y_p + h^2 V_n Y_n\right) \tag{A.11}$$

$$I_1 = \frac{1}{\sqrt{3}}\left(h V_p + h^2 V_n\right) Y_1 \tag{A.12}$$

$$I_2 = \frac{1}{\sqrt{3}}\left(h^2 V_p + h V_n\right) Y_2. \tag{A.13}$$

Substituting (A.9)–(A.13) into (3.5) and rearranging terms, one obtains:

$$V_n = V_p . \frac{h Y_1 - h^2 Y_2 - (h^2 - h) Y_p}{-h^2 Y_1 + h Y_2 + (h - h^2) Y_n}. \tag{A.14}$$

The terminal voltage V is given by

$$V = V_A = \frac{1}{\sqrt{3}}\left(V_p + V_n\right). \tag{A.15}$$

From (A.14) and (A.15),

$$V_p = \sqrt{3}\, V . \frac{Y_n - \frac{h}{1-h} Y_1 + \frac{1}{1-h} Y_2}{Y_1 + Y_2 + Y_p + Y_n} \tag{A.16}$$

or

$$V_p = \sqrt{3}\, V . \frac{Y_n + \frac{e^{-j\pi/6}}{\sqrt{3}} Y_1 + \frac{e^{j\pi/6}}{\sqrt{3}} Y_2}{Y_1 + Y_2 + Y_p + Y_n}. \tag{3.6}$$

Substituting (A.16) into (A.14),

$$V_n = \sqrt{3}\, V \cdot \frac{Y_p + \frac{1}{1-h}Y_1 - \frac{h}{1-h}Y_2}{Y_1 + Y_2 + Y_p + Y_n} \tag{A.17}$$

or

$$V_n = \sqrt{3}\, V \cdot \frac{Y_p + \frac{e^{j\pi/6}}{\sqrt{3}}Y_1 + \frac{e^{-j\pi/6}}{\sqrt{3}}Y_2}{Y_1 + Y_2 + Y_p + Y_n}. \tag{3.7}$$

A.4 Derivation of Angular Relationship

Applying the sine rule to the current phasor triangle in Figure 3.7(a),

$$\frac{I_1}{\sin\theta_1} = \frac{I_{L2}}{\sin\theta_2}$$

or

$$\frac{I_1}{\sin(2\pi/3 - \phi_p)} = \frac{I_{LINE}}{\sin(2\pi/3)}$$

where I_{LINE} is the line current of the IG under perfect phase balance. Therefore,

$$I_1 = \frac{2I_{LINE}}{\sqrt{3}} \sin(2\pi/3 - \phi_p). \tag{A.18}$$

Applying the sine rule to the current phasor triangle in Figure 3.7(b),

$$\frac{I}{\sin\theta_5} = \frac{I_1}{\sin\gamma}$$

or

$$\frac{I}{\sin(2\pi/3 - \phi_p)} = \frac{I_1}{\sin\gamma}$$

Therefore,

$$\sin\gamma = \frac{I_1}{I} \sin(2\pi/3 - \phi_p). \tag{A.19}$$

Applying the cosine rule to the current phasor triangle in Figure 3.7(b),

$$I = \sqrt{I_{L3}^2 + I_1^2 - 2I_{L3}I_1 \cos\theta_5}$$

$$= \sqrt{I_{LINE}^2 + \frac{4}{3}I_{LINE}^2 \sin^2(2\pi/3 - \phi_p) - 2I_{LINE}^2 \sin(2\pi/3 - \phi_p).\cos(2\pi/3 - \phi_p)}$$

$$= \frac{I_{LINE}}{\sqrt{3}} \cdot \sqrt{3 + 4\sin^2(2\pi/3 - \phi_p) - 2\sqrt{3}\sin 2(2\pi/3 - \phi_p)}. \tag{A.20}$$

Substituting (A.18) and (A.20) into (A.19),

$$\sin \gamma = \frac{2 \sin^2(2\pi/3 - \phi_p)}{\sqrt{3 + 4 \sin^2(2\pi/3 - \phi_p) - 2\sqrt{3} \sin 2(2\pi/3 - \phi_p)}} \tag{A.21}$$

If we let $\alpha = 2\pi/3 - \phi_p$, then (A.21) is simplified to

$$\sin \gamma = \frac{2 \sin^2 \alpha}{\sqrt{3 + 4 \sin^2 \alpha - 2\sqrt{3} \sin 2\alpha}}. \tag{3.12}$$

A.5   Input Impedance of SEIG with the Steinmetz Connection

From (5.2), it is obvious that (A.8) is also valid for the three-phase SEIG with the Steinmetz connection as shown in Figure 5.1. The phase currents are also given by (A.9) and (A.11), while the terminal voltage V is given by (A.15).

Substituting (A.6)–(A.11) into (5.3), one obtains

$$h^2 \frac{V_p}{Z_C} + h \frac{V_n}{Z_C} = (h - h^2) \frac{V_p}{Z_p} + (h^2 - h) \frac{V_n}{Z_n}$$

or

$$V_n = V_p \cdot \frac{Z_C + \frac{e^{-j\pi/6}}{\sqrt{3}} Z_p}{Z_C + \frac{e^{j\pi/6}}{\sqrt{3}} Z_n} \cdot \frac{Z_n}{Z_p}. \tag{A.22}$$

Substituting (A.22) into (A.15),

$$V_p = \sqrt{3} V \cdot \frac{Z_p \left(Z_C + \frac{Z_n}{1-h} \right)}{Z_p Z_n + Z_p Z_C + Z_n Z_C} \tag{A.23}$$

or

$$V_p = \sqrt{3} V \cdot \frac{Z_p \left(Z_C + \frac{e^{j\pi/6}}{\sqrt{3}} Z_n \right)}{Z_p Z_n + Z_p Z_C + Z_n Z_C}. \tag{5.6}$$

Substituting (A.23) into (A.22),

$$V_n = \sqrt{3} V \cdot \frac{Z_n \left(Z_C + \frac{e^{-j\pi/6}}{\sqrt{3}} Z_p \right)}{Z_p Z_n + Z_p Z_C + Z_n Z_C}. \tag{5.7}$$

From (5.6) and (5.7)

$$I_p = \frac{V_p}{Z_p} = \sqrt{3}V.\frac{Z_C + \frac{e^{j\pi/6}}{\sqrt{3}}Z_n}{Z_p Z_n + Z_p Z_C + Z_n Z_C} \tag{A.24}$$

$$I_n = \frac{V_n}{Z_n} = \sqrt{3}V.\frac{Z_C + \frac{e^{-j\pi/6}}{\sqrt{3}}Z_p}{Z_p Z_n + Z_p Z_C + Z_n Z_C}. \tag{A.25}$$

From (5.4) and (A.2), the input current I is given by

$$I = I_A - I_C = \frac{1}{\sqrt{3}}\left[(1 - h)I_p + (1 \quad h^2)I_n\right]. \tag{A.26}$$

Substituting (A.24) and (A.25) into (A.26) and simplifying,

$$I = V.\frac{3Z_C + Z_p + Z_n}{Z_p Z_n + Z_p Z_C + Z_n Z_C}. \tag{A.27}$$

When viewed across terminals 1 and 3, the input impedance of the SEIG in Figure 5.1 is given by

$$Z_{in} = \frac{V}{I} = \frac{Z_p Z_n + Z_p Z_C + Z_n Z_C}{3Z_C + Z_p + Z_n}. \tag{5.8}$$

References

[1] J.E. Brown and O.I. Butler, 'A general method of analysis of 3-phase induction motors with asymmetrical primary connections', *IEE Proceedings*, Pt II, Vol. 100, pp. 25–34, February 1953.
[2] C.F. Wagner and R.D. Evans, *Symmetrical* Components, McGraw-Hill, New York, 1933.
[3] M.G. Say, *Alternating Current Machines*, 5th Edn, pp. 333–336, Pitman (ELBS), London, 1983.

Appendix B
The Method of Hooke and Jeeves

The pattern search method of Hooke and Jeeves [1] consists of a sequence of *exploratory moves* about a base point which, if successful, are followed by *pattern moves*. The procedure is summarized as follows.

Exploratory moves. The purpose of an exploratory move is to acquire information about the function $f(x)$ in the neighbourhood of the current base point b_k. Each variable x_i, in turn, is given an increment ε_i (first in the positive direction and then, if necessary, in the negative direction) and a check is made of the new function value. If any move is a success (i.e. results in a reduced function value), the new value of that variable will be retained. After all the variables have been considered, a new base point b_{k+1} will be reached. If $b_{k+1} = b_k$, no function reduction has been achieved. The step length ε_i is reduced (say to 1/10th of its current value) and the procedure is repeated. If $b_{k+1} \neq b_k$, a pattern move from b_k is made.

Pattern moves. A pattern move attempts to speed up the search by using the information already acquired about $f(x)$ so as to identify the best search direction. By intuition, a move is made from b_{k+1} in the direction $b_{k+1} - b_k$, since a move in this direction has led to a decrease in the function value. Thus, the function is evaluated at the next pattern point given by

$$p_k = b_k + 2(b_{k+1} - b_k). \tag{B.1}$$

The search continues with a new sequence of exploratory moves about p_k. If the lowest function value obtained is less than $f(b_k)$, then a new base point b_{k+2} has been reached. In this case a second pattern move is made (using (B.1) with all suffixes increased by unity). If not, the pattern move from b_{k+1} is abandoned and we continue with a new sequence of exploratory moves about b_{k+1}.

Distributed Generation: Induction and Permanent Magnet Generators   L. L. Lai and T. F. Chan
© 2007 John Wiley & Sons, Ltd

The minimum is assumed to be obtained if the step length for each variable has been reduced to a specified small value.

For the SEIG problem, the two variables are the per-unit frequency a and the magnetizing reactance X_m. After a successful voltage build-up, a must be less than the per-unit speed b and X_m must be less than the unsaturated magnetizing reactance X_{mu}. To start the Hooke and Jeeves pattern search procedure, initial estimates of a and X_m can be chosen to be b and X_{mu}, respectively. For small load impedances, however, it was found that an initial value of $0.97b$ for a would give better convergence performance.

In the computer programs, an initial step length of 0.01 was chosen for both variables and convergence was assumed to be obtained if the step length was reduced to less than 1.0e-8. All the circuit parameters used were expressed in per-unit values.

Reference

[1] Byron S. Gottfried and Joel Weisman, *Introduction to Optimization Theory*, Prentice Hall, Englewood Cliffs, NJ, 1973.

Appendix C

A Note on the Finite Element Method [1]

C.1 Energy Functional and Discretization

The solution of field problems using the finite element method involves the minimization of an energy functional, since the total energy in any physical system should be a minimum under steady-state conditions. The two-dimensional field problem is formulated in terms of magnetic vector potential A and the coil currents (and hence the current densities J) are assumed to be known. The appropriate energy functional to be minimized is

$$E(A) = \iint_\Omega \left(\int_0^B \frac{B}{\mu} .dB - A.J \right) dx.dy. \tag{4.3}$$

It will be assumed that the permeability μ at any point is constant at the value corresponding to the final flux density B. Equation (4.3) can then be written as

$$E(A) = \iint_\Omega \left(\frac{B^2}{2\mu} - J.A \right) dx.dy = \iint_\Omega \left\{ \frac{1}{2\mu} \left[\left(\frac{\partial A}{\partial x} \right)^2 + \left(\frac{\partial A}{\partial y} \right)^2 \right] - J.A \right\} dx.dy. \tag{C.1}$$

Consider the discretized solution region (Figure 4.7) with n_e elements and n_p nodes. Because the energy of the functional is a scalar quantity, the energy in the global field region is considered to be the sum of energies in individual elements. Therefore, we

Distributed Generation: Induction and Permanent Magnet Generators   L. L. Lai and T. F. Chan
© 2007 John Wiley & Sons, Ltd

aim to find the nodal potentials $A_1, A_2, \ldots, A_{n_p}$, such that the following condition is satisfied:

$$E(A) = E(A_1, A_2, \ldots, A_{n_p}) = \min. \tag{C.2}$$

That is,

$$\frac{\partial E(A)}{\partial A_p} = 0 \qquad (p = 1, 2, 3, \ldots, n_p) \tag{C.3}$$

or

$$\frac{\partial E_1(A)}{\partial A_p} + \frac{\partial E_2(A)}{\partial A_p} = 0 \qquad (p = 1, 2, 3, \ldots, n_p) \tag{C.4}$$

where

$$E_1(A) = \iint_\Omega \left\{ \frac{1}{2\mu} \left[\left(\frac{\partial A}{\partial x} \right)^2 + \left(\frac{\partial A}{\partial y} \right)^2 \right] \right\} dx.dy \tag{C.5}$$

$$E_2(A) = - \iint_\Omega J.A \, dx.dy. \tag{C.6}$$

C.2   Shape Functions

Assume that the solution region has been divided into a number of triangular elements, or subregions, as shown in Figure C.1. For a typical element e, the vertices i, j and m are assigned in the counterclockwise sense. The magnetic vector

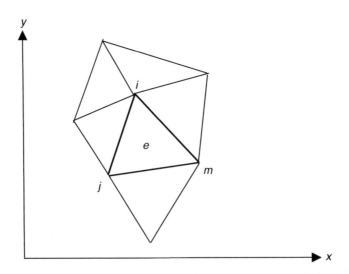

Figure C.1   *Discretization of field region by triangular elements*

potential at any point (x, y) within element e is approximated by the following linear interpolating function:

$$A(x, y) = \beta_1 + \beta_2 x + \beta_3 y \qquad (C.7)$$

where β_1, β_2 and β_3 are the coefficients to be determined.

If the coordinates of the vertices i, j and m are (x_i, y_i), (x_j, y_j) and (x_m, y_m), while the magnetic vector potentials are A_i, A_j and A_m, respectively, then from (C.7)

$$
\begin{aligned}
A_i &= \beta_1 + \beta_2 x_i + \beta_3 y_i \\
A_j &= \beta_1 + \beta_2 x_j + \beta_3 y_j \\
A_m &= \beta_1 + \beta_2 x_m + \beta_3 y_m.
\end{aligned}
\qquad (C.8)
$$

The coefficients can be determined by solving (C.8):

$$
\begin{aligned}
\beta_1 &= \frac{1}{2\Delta_e}(a_i A_i + a_j A_j + a_m A_m) \\[2mm]
\beta_2 &= \frac{1}{2\Delta_e}(b_i A_i + b_j A_j + b_m A_m) \\[2mm]
\beta_3 &= \frac{1}{2\Delta_e}(c_i A_i + c_j A_j + c_m A_m)
\end{aligned}
\qquad (C.9)
$$

where

$$
\begin{aligned}
a_i &= x_j y_m - x_m y_j; & a_j &= x_m y_i - x_i y_m; & a_m &= x_i y_j - x_j y_i \\
b_i &= y_j - y_m; & b_j &= y_m - y_i; & b_m &= y_i - y_j \\
c_i &= x_m - x_j; & c_j &= x_i - x_m; & c_m &= x_j - x_i
\end{aligned}
\qquad (C.10)
$$

and Δ_e denotes the area of the triangular element e, i.e.

$$\Delta_e = \frac{1}{2}\begin{vmatrix} 1 & x_i & y_i \\ 1 & x_j & y_j \\ 1 & x_m & y_m \end{vmatrix} = \frac{1}{2}(b_i c_j - b_j c_i) = \frac{1}{2}(a_i + a_j + a_m). \qquad (C.11)$$

Substituting (C.9) into (C.7) and rearranging terms,

$$
\begin{aligned}
A(x, y) = \frac{1}{2\Delta_e}&[(a_i + b_i x + c_i y)A_i + (a_j + b_j x + c_j y)A_j \\
&+ (a_m + b_m x + c_m y)A_m]
\end{aligned}
\qquad (C.12)
$$

or

$$A(x, y) = N_i A_i + N_j A_j + N_m A_m \qquad (C.13)$$

where

$$N_i = \frac{1}{2\Delta_e}(a_i + b_i x + c_i y)$$

$$N_j = \frac{1}{2\Delta_e}(a_j + b_j x + c_j y) \tag{C.14}$$

$$N_m = \frac{1}{2\Delta_e}(a_m + b_m x + c_m y).$$

N_i, N_j and N_m are known as the shape functions and they all depend on the geometry of the triangular element e.

Taking partial derivatives of (C.12),

$$\frac{\partial A}{\partial x} = \frac{1}{2\Delta_e}(b_i A_i + b_j A_j + b_m A_m)$$

$$\frac{\partial A}{\partial y} = \frac{1}{2\Delta_e}(c_i A_i + c_j A_j + c_m A_m). \tag{C.15}$$

Consider first the contribution E_{e1} of a typical element e to the total energy functional E_1 in (C.5):

$$E_{e1}(A) = \iint_{\Omega_e} \left\{ \frac{1}{2\mu} \left[\left(\frac{\partial A}{\partial x} \right)^2 + \left(\frac{\partial A}{\partial y} \right)^2 \right] \right\} dx.dy. \tag{C.16}$$

The terms in square brackets in (C.16) can be expressed as

$$\left(\frac{\partial A}{\partial x} \right)^2 + \left(\frac{\partial A}{\partial y} \right)^2 = \begin{bmatrix} \dfrac{\partial A}{\partial x} & \dfrac{\partial A}{\partial y} \end{bmatrix} \begin{bmatrix} \dfrac{\partial A}{\partial x} \\ \dfrac{\partial A}{\partial y} \end{bmatrix}$$

$$= \frac{1}{4\Delta_e^2} \begin{bmatrix} A_i & A_j & A_m \end{bmatrix} \begin{bmatrix} b_i & c_i \\ b_j & c_j \\ b_m & c_m \end{bmatrix} \begin{bmatrix} b_i & b_j & b_m \\ c_i & c_j & c_m \end{bmatrix} \begin{bmatrix} A_i \\ A_j \\ A_m \end{bmatrix}. \tag{C.17}$$

But

$$\iint_{\Omega_e} dx.dy = \Delta_e \tag{C.18}$$

therefore (C.16) may be written as

$$E_{e1}(A) = \frac{1}{4\mu\Delta_e} \cdot \frac{1}{2} \begin{bmatrix} A_i & A_j & A_m \end{bmatrix} \begin{bmatrix} b_i^2 + c_i^2 & b_i b_j + c_i c_j & b_i b_m + c_i c_m \\ b_j b_i + c_j c_i & b_j^2 + c_j^2 & b_j b_m + c_j c_m \\ b_m b_i + c_m c_i & b_m b_j + c_m c_j & b_m^2 + c_m^2 \end{bmatrix} \begin{bmatrix} A_i \\ A_j \\ A_m \end{bmatrix}$$

$$= \frac{1}{2} \begin{bmatrix} A_i & A_j & A_m \end{bmatrix} \begin{bmatrix} K_{ii}^e & K_{ij}^e & K_{im}^e \\ K_{ji}^e & K_{jj}^e & K_{jm}^e \\ K_{mi}^e & K_{mj}^e & K_{mm}^e \end{bmatrix} \begin{bmatrix} A_i \\ A_j \\ A_m \end{bmatrix}$$

$$= \frac{1}{2} [A]_e^T [K]_e [A]_e \tag{C.19}$$

where

$$[A]_e = \begin{bmatrix} A_i \\ A_j \\ A_m \end{bmatrix}$$

and $[A]_e^T$ is the transpose of $[A]_e$.

The elements of matrix $[K]_e$ are given as follows:

$$K_{st}^e = \frac{1}{4\mu\Delta_e}(b_s b_t + c_s c_t) \quad (s = i, j, m; \quad t = i, j, m). \tag{C.20}$$

C.3 Functional Minimization and Global Assembly

If the global magnetic potential vector is denoted by $[A]$, then the local magnetic potential vector $[A]_e$ of element e, when expressed in the global system, is given by

$$[A]_e = [C]_e [A] \tag{C.21}$$

where $[C]_e$ is known as the connection matrix.

From (C.19), the energy functional $E_{e1}(A)$ can be expressed as

$$E_{e1}(A) = \frac{1}{2} [A]^T \left([C]_e^T [K]_e [C]_e \right) [A] = \frac{1}{2} [A] T [K^*]_e [A] \tag{C.22}$$

where

$$[K^*]_e = [C]_e^T [K]_e [C]_e . \tag{C.23}$$

It should be noted that $[K^*]_e$ is an $(n_e \times n_e)$ matrix in which the elements are given by:

$$K_{st}^{*e} = \frac{1}{4\mu\Delta_e}(b_s b_t + c_s c_t) \quad (s = i, j, m; \quad t = i, j, m) \qquad \text{(C.24a)}$$

$$K_{st}^{*e} = 0 \quad (s \neq i, j, m; t \neq i, j, m). \qquad \text{(C.24b)}$$

The total energy of the field region Ω can now be expressed as

$$E_1(A) = \frac{1}{2}[A]^T \left(\sum_{e=1}^{n_e} [K^*]_e \right) [A] = \frac{1}{2}[A]^T [K][A] \qquad \text{(C.25)}$$

where

$$[K] = \sum_{e=1}^{n_e} [K^*]_e. \qquad \text{(C.26)}$$

$[K]$ is known as the coefficient matrix, or the stiffness matrix.

From (C.25),

$$\frac{\partial E_1(A)}{\partial A_p} = [K][A]. \qquad \text{(C.27)}$$

Next, consider the second derivative in (C.4). For element e,

$$E_{e2}(A) = -\iint_{\Omega_e} J.A \, dx.dy. \qquad \text{(C.28)}$$

With J constant within the element, it can be shown that

$$x\frac{\partial E_{e2}(A)}{\partial A_k} = -\iint_{\Omega_e} J.(a_k + b_k x + c_k y)dx.dy = -\frac{J.\Delta_e}{3} \quad (k = i, j, m). \quad \text{(C.29)}$$

Considering the contributions from all the nodes,

$$\frac{\partial E_{e2}(A)}{\partial A_p} = -[R] \qquad \text{(C.30)}$$

where $[R]$ is an $(n_p \times 1)$ column vector in which all the terms are known.

From (C.4), (C.26) and (C.29), the following matrix equation is obtained:

$$[K][A] = [R]. \qquad \text{(4.4)}$$

The right-hand-side vector $[R]$ is also known as the forcing function.

Reference

S.J. Salon, *Finite Element Analysis of Electric Machines*, Kluwer Academic, Boston, 1995.

Appendix D
Technical Data of Experimental Machines

D.1 Machine IG1

Three-phase, delta-connected, 2.2 kW, 220 V, 9.4 A, four-pole, 50 Hz, cage-type induction machine (Sections 3.2, 3.3, 5.2, 5.3, 5.4 and 5.5). The machine parameters are (per-unit values given in brackets)

Stator resistance R_1	$= 3.44\,\Omega$ (0.0844);
Stator leakage reactance X_1	$= 4.56\,\Omega$ (0.112);
Positive-sequence rotor resistance R_{2p}	$= 2.53\,\Omega$ (0.0621);
Negative-sequence rotor resistance R_{2n}	$= 4.0\,\Omega$ (0.0981);
Rotor leakage reactance X_2	$= 4.07\,\Omega$ (0.1);
Core loss resistance R_c	$= 896\,\Omega$ (22.0);
Magnetizing reactance at nominal voltage	$= 71\,\Omega$ (1.74);
Friction and windage loss P_{fw}	$= 47\,\text{W}$ (0.013);
Stray-load loss	$= 1.8\%$ of rated power.

The variation of positive-sequence air gap voltage E_1 with magnetizing reactance X_m is modelled by the following describing equations expressed per unit:

$$E_1 = \begin{cases} 1.345 - 0.203X_m, & X_m < 1.728 \\ 1.901 - 0.525X_m, & 1.728 \leq X_m < 2.259 \\ 3.156 - 1.08X_m, & 2.259 \leq X_m < 2.446 \\ 37.49 - 15.12X_m, & 2.446 \leq X_m < 2.48 \\ 0, & 2.48 \leq X_m. \end{cases} \tag{D.1}$$

Distributed Generation: Induction and Permanent Magnet Generators L. L. Lai and T. F. Chan
© 2007 John Wiley & Sons, Ltd

D.2   Machine IG2

Three-phase, delta-connected, 2.2 kW, 220 V, 9.4 A, 50 Hz, four-pole, cage-type induction machine (Chapter 4). The machine parameters are as follows.

Stator:

Outer diameter	= 155 mm;
Inner diameter	= 98 mm;
Axial length	= 101 mm;
Number of slots	= 36;
Turns per coil	= 43;
Coil span	= 9 slots;
Number of parallel paths	= 1.

Rotor:

Outer diameter	= 97.4 mm;
Inner diameter	= 38 mm;
Number of slots	= 32;
End-ring resistance	= 0.1817e-4 Ω (per layer);
End-ring leakage inductance	= 0.5784e-7 H (per layer).

D.3   Prototype PMSG with Inset Rotor

This machine is used for the investigations in Chapter 7. The specifications of the generator are as follows.

General:

Three-phase, four-pole, 1500 r/min, star-connected, 110 V, 2.5 kVA.

Stator:

Outer diameter	= 155 mm;
Inner diameter	= 98 mm;
Number of slots	= 36;
Coil span	= 7 slots;
Axial core length	= 100 mm;
Turns per coil	= 6;
Cross-sectional area of conductors	= 2.65 mm^2;
Armature resistance per phase	= 0.295 Ω.

Rotor:

Permanent magnet material	= Nd–Fe–B;
Outer diameter of soft-iron pole pieces	= 97.3 mm;
Outer diameter of magnets	= 96.4 mm;

Inner diameter of magnets	= 87.4 mm;
Inner diameter of rotor core	= 38 mm;
Average pole arc of magnets	= 144°e;
Average arc of interpolar soft-iron pole pieces	= 12°e;
Space between magnet and soft-iron pole piece	= 6°e;
Remanence of Nd–Fe–B magnets	= 1.128 T;
Coercive force of Nd–Fe–B magnet	= 880 kA/m.

Index

Distributed Generation: Induction and Permanent Magnet Generators L. L. Lai and T. F. Chan
© 2007 John Wiley & Sons, Ltd

Printed and bound by CPI Group (UK) Ltd, Croydon, CR0 4YY

17/04/2025

14658867-0001